INDUSTRIAL FIRE BRIGADES TRAINING MANUAL

Edited by
CHARLES A. TUCK, JR.

1978

NATIONAL FIRE PROTECTION ASSOCIATION
470 Atlantic Avenue, Boston, Massachusetts 02210

INDUSTRIAL FIRE BRIGADES TRAINING MANUAL

Fifth Edition

NFPA No. SPP-13A
Library of Congress Card No. 77–95083
ISBN 0-87765-116-7

Printed in U.S.A.

TABLE OF CONTENTS

FOREWORD

The first edition of this manual was published in 1943. It was edited by a subcommittee of the NFPA Committee on Firemen's Training, which consisted of Emmet T. Cox, W. Fred Heisler, and Horatio Bond, then NFPA Chief Engineer, and approved by the full committee. The manual was revised by the committee in 1948. In 1954, Mr. Bond and Warren Y. Kimball, then Manager of the NFPA Fire Service Department collaborated to produce the third edition. Mr. Bond edited the fourth edition, which was published in 1968.

The scope of this edition, the fifth, is the same as earlier editions. Material has been revised to reflect advances in the state of the art since the publication of the fourth edition. Though some devices may no longer be manufactured, particularly certain portable fire extinguishers, they are included in this manual because many of them are still in use. However, the emphasis is on modern equipment and techniques.

We have attempted to provide illustrations of equipment representative of that which will be found in the field, but appropriate devices are not limited to those depicted in this manual. We are indebted to the equipment manufacturers who have furnished photographs to illustrate this manual.

We gratefully acknowledge the comments and suggestions of the following members of the Executive Board of the Industrial Fire Protection Section — Robert C. Davis, Chevrolet Division, General Motors Corporation; Bruce Gray, West Point-Pepperell, Inc.; G. G. Robinson, International Harvester Company; Peter K. Schontag, Kaiser Aluminum and Chemical Corporation; John H. Uliana, Bethlehem Steel Corporation; Benjamin F. Vilbert, Getty Oil Company; Donald M. Yarlas, ICI United States.

Charles A. Tuck, Jr.
Editor

December 1977

FOREWORD

INTRODUCTION

The management of an industrial property should have some procedure and organization for coping with fire and related emergencies. Often the plant is small, and total reliance is placed upon the community fire department. Where this is the case, management or its representative merely has to arrive at a good working arrangement with the public fire chief to ensure that fire department operations on the premises will be the most effective possible. In other cases, the plant protection organization might consist of first aid fire fighting teams, or, in large properties, it may be a fully equipped private fire department.

Though some operating functions that may have to be performed during an emergency may have no counterpart in the normal daily routine, it is not necessary to create a comprehensive organization to direct general operations during an emergency. However, the plant manager should modify plant operations and provide additional equipment and personnel that may be needed to meet various emergencies.

A special organization that is to be mobilized to direct general operations during times of emergency should, as much as possible, be consistent with the normal operating organization. For example, if a critical process is involved and has to be shut down, it should be shut down by qualified technical personnel familiar with the process. The medical department should be responsible for establishing first aid stations, and the public relations department should deal with the news media. The guard force should maintain security and control traffic, both pedestrian and vehicular, on the premises.

Though it recognizes the importance of emergency support functions, such as those mentioned in the preceding paragraphs, this manual is concerned primarily with the organization, training, and operations of the plant fire fighting force.

Some chapters may go into greater detail than would be considered necessary for first aid fire fighting teams, but the instructor has the option of selecting those chapters which are applicable to plant organization and conditions. The instructor may wish to supplement this manual with additional information on operational practices obtained from fire department drill schools.

Chapter 1

ORGANIZATION and TRAINING

The management of every industrial property should provide, equip, and train an organization to deal with fires and related emergencies. The size of the facility, the presence of unusual hazards, the potential magnitude of a fire within the property or of an exposure fire, and the availability of a public or private fire department will help to determine the nature of the industrial fire protection organization to be provided. Public or private fire departments do not necessarily take the place of the industrial fire brigade in a large property.

ORGANIZATION

In its simplest form, this organization would consist of the manager of the property assisted by selected personnel. Most plants would have only first-aid fire fighters who would be under the direction of their usual supervisors. In properties where more persons are available, they should be organized as a team, or teams, to function as a private fire brigade. Individual fire brigades may respond to alarms in all areas of a property, or each geographical or functional area may have a separate fire brigade organization according to the needs of the property. The organization should be such that a fire brigade is on duty on each working shift and at periods when the plant is shut down.

The equipment that must be put into service at a fire will determine the number of persons required for each operating unit or company into which the brigade is organized and the total number of people needed in the brigade. Operating units or companies may be composed of two or more members to operate a specific item of equipment or of a larger group to perform more complicated operations. Each company should have a leader, and each brigade should have a chief.

Another form of plant fire protection organization would include employees who devote all of their time to fire protection matters, supplemented by additional employees who would be released from their normal duties for fire department duty when the need arises.

FIRE BRIGADE CHIEFS

The property manager should be accountable for the prevention of fire loss. The manager may assign a fire loss prevention specialist to perform the duties directly associated with the manager's responsibility for loss prevention. From the duties enumerated for each, note that the functions of the loss prevention specialist and the duties of the chief of a fire brigade are not the same.

The loss prevention specialist acting for the manager or the manager himself should:

(*a*) Provide equipment and supplies for the fire brigade or brigades.

(*b*) Establish the size and organizational structure of the fire brigades.

(*c*) See that the brigades are suitably staffed and trained.

(*d*) Select the fire brigade chiefs.

Fire brigade chiefs should have administrative and supervisory abilities. A fire brigade chief should have duties including the following:

(a) Periodic evaluation of the equipment provided for fire fighting. The chief should be responsible for setting in motion necessary procedures for replacing missing equipment or for repairing inoperative equipment. The chief should also call to the immediate attention of the property manager or the loss prevention specialist any situation that is likely to reduce the effectiveness of fire fighting operations.

(b) Provision of plans of action to meet possible fire situations in the plant, subject to the approval of the manager and the fire loss prevention specialist.

(c) Periodic review of the brigade roster and preparation of recommendations that additional members be selected, appointed, and made available to keep up the roster.

(d) Preparing the plan for training members of the brigade and other employees.

Enough assistant chiefs should be appointed to cover the chief's position around the clock. Their rank, one to another and to the chief, should be established to provide for succession in event of absence.

MEMBERSHIP OF THE FIRE BRIGADE

Members of the fire brigade should be persons who have met qualifications appropriate for fire brigade work at the particular property. Brigade membership should consist of the necessary personnel for fire fighting teams and include certain operating and maintenance personnel. The property manager should identify the persons who have to perform duties to assist brigade members in an emergency by reason of their normal responsibilities and decide which ones should be assigned these functions. (Electricians are an example.)

To qualify as a member of the fire brigade, individuals should be available to answer alarms and to attend required training sessions. A prearranged schedule for availability should be established to prevent conflict of duties and to cover absences, such as regular off-duty periods, vacations, and sickness.

Minimum physical requirements should be established. A periodic physical examination is desirable.

Members of the brigade should be given some appropriate identification that will be recognized by fire and police officials as well as by plant guards. Such identification, which may be in the form of a card or badge, for example, will assist brigade members in reaching the plant in an emergency and facilitate their movement throughout the plant to where their services are needed.

TRAINING

A schedule of training should be established for members of the brigade. Members should be required to complete a specified program of instruction as a condition to membership in the brigade. Training sessions should be held at least monthly.

All members of the brigade, regardless of the fire fighting team or company to which they belong, should be trained. They should be instructed in the handling of any and all of the fire and rescue apparatus provided. The training program should be adapted to the purpose of the particular brigade. It should include fire fighting with portable fire extinguishers, the use of hose lines, ventilation of buildings, salvage operations, and performing related rescue operations.

The training program should keep up with problems presented by new fire hazards in the property, with new fire extinguishing equipment, and with methods provided for property protection.

Assistance in setting up and training the fire brigade can be obtained from outside agencies. Among these are the municipal fire departments, state fire schools, state educational extension services, state fire marshals' departments, state insurance inspection bureaus, colleges, and others who give fire service training. Members of the brigade should be afforded opportunities to improve their knowledge of fire fighting and fire prevention through attendance at meetings and special training classes where available.

Where the number of individuals participating in the fire brigade training program warrants such arrangement, a special space or room in the property for fire brigade use should be available for that part of a training program requiring lectures or classroom instruction. Training aids such as books, literature, and films should be kept at such a location when one is provided.

Practice drills should be held to check the ability of members to perform the operations they are expected to carry out with the fire equipment provided. Drills should occasionally be held under adverse weather conditions to work out special procedures needed under these conditions.

In drills, equipment should be operated whenever possible. For example, portable extinguishers should be actually discharged, respiratory protective equipment should be operated, and hose lines should be charged with water.

Practice drills should always be carried out under the control of the chief and leaders of companies at a moderate pace with emphasis on effectiveness rather than speed. This is to assure proper technique and safe operation as required at a fire.

At the conclusion of practice drill, equipment should be promptly placed in readiness to respond to a fire call.

MUTUAL AID

In highly industrialized areas, especially where a single industry predominates, it is not unusual to find an industrial mutual aid organization. The purpose of such an organization is to provide assistance to a member plant when its fire fighting resources are heavily taxed or when special equipment is required.

There are a number of precautions that should be taken in connection with a mutual aid program. Before a plant enters into such an agreement, the company's legal department should approve its terms and adequate liability insurance should be in force. Once the agreement has been made, fire brigade members of participating plants should be made aware of their obligations and trained to handle the special hazards they may encounter in other plants. Each member company should determine what equipment and personnel it can commit to the program without jeopardizing its own plant. A master list of manpower and equipment should be developed and kept at a central location, which would also serve as a dispatching center. The municipal fire department should be advised of the existence of the mutual aid agreement and invited to join it. Whenever it becomes necessary to activate the mutual aid program, nothing should delay notification of the public fire department that a fire is in progress.

EQUIPMENT

The brigade should be provided with equipment and tools of a variety and in such numbers as to enable it to perform the service for which it is organized. This equipment should include items additional to the fixed or portable equipment provided in buildings and yards.

The property manager, the fire loss prevention specialist, or the fire brigade chief should maintain a list of equipment available in the property that might be useful in fire brigade work but is not in the custody of the fire brigade. The list should show where each item is usually located and the name of the department or person in whose custody it may be found. A list of equipment and service agencies from which equipment or assistance may be needed, together with phone numbers, should be kept up to date.

The specific equipment supplied for fire brigade use will depend on the nature of the occupancy and its hazards and on the operations the fire brigade is expected to perform. Equipment categories that should be considered when outfitting a fire brigade include portable fire extinguishers, hose and hose accessories, portable lighting equipment, forcible entry tools, ladders, salvage equipment, rescue and first-aid equipment, personnel protective equipment, transportation facilities, and spare and replacement equipment. The latter category includes equipment that fire brigade members might be called upon to replace — for example, fusible links for fire doors and automatic sprinklers.

INSPECTION AND MAINTENANCE

Certain duties in the periodic inspection and maintenance of plant fire equipment, both fixed and portable, may be assigned to members of the fire brigade. However, the plant manager or the fire loss prevention manager should establish the necessary schedules for such work and should assign these duties to specific personnel and see that these inspections and maintenance operations are carried out and reports filed with management.

Large plants should consider maintaining inspectors, who are not necessarily members of the fire brigade, to carry on a continuous inspection and service program for fire protection equipment.

PUBLIC FIRE DEPARTMENT

It is essential that a clear understanding exist between the public fire department and the plant fire protection organization. Certain processes and materials may be so complicated that the public fire department could not be expected to know how to deal with them. In such cases, the procedures needed to control a fire may not be the conventional ones. It may be necessary to shut down a process or transfer material from one location to another through piping systems. When these conditions exist, certain plant personnel may be much more qualified than the public fire department to handle the situation. It is important, therefore, that all parties concerned agree on who is to direct operations in the event a fire does occur.

PREFIRE PLANNING

Prefire planning is the mapping out of steps to be taken against a probable fire. The action agreed upon is based on the experiences of those participating in the planning process, known or existing conditions, cause and effect, and reasonable expectancy of a fire occurring. There are basically four steps to be followed in prefire planning — the collection of all information that may be pertinent to fire fighting operations at a given location, analysis of that information, dissemination of the information to all concerned, and classroom review and drill. The prefire planning process should include members of the mutual aid organization and the public fire department.

Suggested Reading

NFPA 6, *Recommendations for Organization of Industrial Fire Loss Prevention*, 1974, NFPA, Boston.

NFPA 7, *Recommendations for Management Control of Fire Emergencies*, 1974, NFPA, Boston.

NFPA 8, *Management Responsibility for Effects of Fire on Operations*, 1974, NFPA, Boston.

Fire Protection Handbook, 14th Edition, 1976, NFPA Boston, Section 10.

NFPA Inspection Manual, 4th Edition, 1976, NFPA, Boston, Chapter 2.

Instructor Training, 3rd Edition, 1973, International Fire Service Training Association, Stillwater, Oklahoma.

Chapter 2

FIRE HAZARDS

In order to perform effectively, industrial fire fighters should know something about fire — what it is, its causes, the conditions under which it may be expected to occur, and its extinguishment. They should learn to recognize situations that are potential fire hazards.

UNDERSTANDING FIRE

Generally, fire is the result of the chemical combination of a combustible material (fuel) with oxygen in the presence of heat. Usually, the process is accompanied by the generation of light (glowing or flaming combustion); however, the flame from such materials as hydrogen or alcohol may be difficult, if not impossible, to see in bright daylight.

Though the most common source of oxygen for combustion is the air around us, a few materials have the ability to release oxygen from their own chemical structure in such a way that may support partial combustion without any additional oxygen from an external source.

Not all materials require oxygen in order to burn. For example, iron wire will burn in an atmosphere of chlorine gas, and zirconium dust will burn in carbon dioxide. However, for purposes of discussion, we will consider the most common form of combustion — that supported by oxygen.

Combustion

The first ingredient in ordinary combustion is a combustible material or fuel. A combustible material contains chemical elements that will react with oxygen, under proper conditions, to produce fire. Carbon is the most common fuel element. Fuel elements less frequently encountered, but nevertheless of considerable concern, are hydrogen and sulfur. Most combustible materials contain one or more of these elements.

Common Combustible Materials

(C — carbon; H — hydrogen; O — oxygen; N — nitrogen; S — sulfur; the number with each indicates the number of atoms of each element.)

Fuel Materials	Elements
Coal	C+Bitumen
Wood, Paper, Cotton	$C_{16}H_{10}O_5$ (approximate)
Silk, Rayon, Wool	$C_{51}H_7O_{21}N_{17}S_4$ (approximate)
Rubber	$C_{10}H_{16}$+other minor elements
Gasoline	C_5H_{12} to C_9H_{20}
Kerosine	$C_{10}H_{22}$ to $C_{15}H_{32}$
Alcohol, Grain	C_2H_5OH
Alcohol, Wood	CH_3OH
Ether	$C_2H_5OC_2H_5$
Naphtha	C_4H_{10} to C_7H_{16}
Acetone	CH_3COCH_3
Turpentine	$C_{10}H_{16}$
Carbon Disulfide	CS_2
Butane (liquid gas)	C_4H_{10}
Propane (bottled gas)	C_3H_8
Methane (natural gas)	CH_4
Phenol (carbolic acid)	C_6H_5OH
Formaldehyde	CH_2O
Acetylene	C_2H_2

A material is not necessarily a combustible material merely because it contains a fuel element. Certain materials are noncombustible because a major portion of their formulation consists of elements that do not ignite readily at ordinary temperatures. One such formulation is bromotrifluoromethane ($BrCF_3$), which contains one atom of carbon, a combustible element, and one atom of bromine, and three atoms of fluorine, both noncombustible elements. As a matter of fact, bromotrifluoromethane, also known as Halon 1301, is used as an extinguishing agent. The proportion of combustible to noncombustible elements in a compound is not the sole factor in determining the combustibility of that compound. The ease with which the elements combine with oxygen and the manner in which they combine must also be considered, along with other factors. However, the proportion of combustible to noncombustible elements will provide a rough idea of the compound's combustibility.

The second factor in ordinary combustion is oxygen. Since air is approximately 21 percent oxygen, that element is present almost everywhere fuels are found. For complete combustion, the three most common fuel elements require oxygen in the following proportions (parts by volume):

Carbon — 1 part	Oxygen — 2 parts
Hydrogen — 2 parts	Oxygen — 1 part
Sulfur — 1 part	Oxygen — 2 parts

Heat is the third factor necessary for combustion to take place. Before fire (oxidation) starts, heat must raise the temperature of the fuel surface to a point where chemical union of the fuel and oxygen occurs.

These three factors are necessary before fire occurs. There must be a combustible material, oxygen, and sufficient heat to start combustion. By breaking any one of the three links of the chain — by eliminating the combustible material, by cutting off the oxygen supply, or by cooling to a point below the ignition temperature — fire can be extinguished.

There is a fourth extinguishment mechanism involving the chemical inhibition of flame by certain halogenated hydrocarbons, alkali metal salts, and ammonium salts. The exact manner in which the extinguishing agent chemically interferes with the combustion chain is not certain. However, the method works rapidly and efficiently on flaming vapor and gaseous fuels. It is not effective against glowing combustion.

Ignition

Before ignition can take place, not only must a source of sufficiently high temperature be present, but also a sufficient quantity of heat to cause ignition must be available. In order for a combustible to ignite, its temperature must be raised to the point where oxidation will occur. Oxidation, itself, produces more heat, and thus, fire spreads. However, if heat is carried away from the ignition point faster than it is produced, temperature will decrease, and the fire will go out. Heat can be carried away by radiation, conduction, and convection, as is explained under **Heat Transfer.**

Consider a match and a wooden log. The log may ignite at the point where the match flame touches it, but heat will be conducted away from the ignition point and absorbed by the body of the log. Some heat will be lost through convection and radiation. As soon as the match flame is removed, heat will be transferred away from the point of ignition faster than it is produced, and the fire will go out. However, if a sufficient quantity of heat were applied to the entire log, the wood would soon become saturated with heat, and little, if any, would be conducted away from the burning surface. The fire would continue to burn.

Though the log may be subjected to a quantity of heat large enough to severely limit loss through conduction, fire in a single log may go out because of the rapid loss of heat by convection and radiation. However, if two or three burning logs are placed close together, radiation from each will tend to keep the others hot, and combustion will continue. The same effect will occur in the case of a fire burning within a hollow wooden partition. Each surface will radiate heat toward the other, producing a more severe fire than a similar amount of wood surface burning in an open room.

Small bits of combustible material have very little heat absorbing capacity; therefore, a smaller quantity of heat will ignite them. For example, a single match can readily ignite ordinary wood shavings. Where the combustible material is still more finely divided, such as in the form of a dust cloud in air, an even smaller heat source can cause ignition. The same is true of flammable gases and vapors, where the combustible material is divided into individual molecules. This explains why a cloud of fine combustible dust or flammable gas or vapor can be ignited by a source producing a very small quantity of heat, such sources as a small brand or a spark of static electricity. The temperature of such sparks is high enough to ignite any ordinary combustible material, but the quantity of heat produced in a small spark, which lasts only an instant, is not sufficient to cause ignition except under extremely favorable conditions, such as are presented by combustible dusts or flammable gases or vapors.

Ignition temperatures of ordinary combustibles lie in the approximate range of 400 to 1,100 degrees Fahrenheit (204 to 593 degrees Celsius). The exact ignition temperature may vary and depends on such factors as particle size, duration of exposure to heat, and moisture content. However, all common combustible materials have ignition temperatures below the temperature of a match flame, a glowing cigarette or brand, an electric arc, or the heat of friction. Thus, the major consideration is not the ignition temperature, but the quantity of heat available.

There are a few materials with abnormally low ignition temperatures that are of special concern. Carbon disulfide, which is used in many industries, is a flammable liquid that can be ignited at the temperature of ordinary steam pipes. Nitrocellulose materials, such as pyroxylin plastics (celluloid), will start to decompose at 300 degrees Fahrenheit (149 degrees Celsius) and will ignite without further application of heat. Some materials, such as sodium and potassium (in the presence of water) and white phosphorous, will ignite when exposed to air at room temperature.

Sources of Ignition

From what sources can sufficient heat for igniting the combustible fuel-oxygen mixture be expected? The most commonly experienced source of heat is from other fire. A match produces a flame with which to ignite the fuel-oxygen mixture from a gas burner. With a match flame, a handful of paper can be ignited. This can, in turn, ignite wood kindling. If gasoline is spilled, it vaporizes at room temperature; and the vapor thus formed, being heavier than air, drifts along the floor. If it comes in contact with an open flame, the requirements for combustion are met and a fire or explosion results.

Many fuels do not require as much heat as a flame producing approximately 1,000 to 2,000 degrees Fahrenheit (538 to 1,093 degrees Celsius) but can be ignited by hot metals or other materials. When we consider that iron must be heated to a temperature of 752 degrees Fahrenheit (400 degrees Celsius) to be barely visible in the dark, it is apparent that gasoline vapors with an ignition temperature of 495 degrees Fahrenheit (257 degrees Celsius) may be ignited by iron that is far from being red hot. Likewise, dry wood, paper, and other cellulosic material can be ignited by contact with materials far below the temperature of flame.

Overheating and arcing (sparking) are the two electrical sources of ignition heat. When electric current flows through a wire or other conductor, that conductor offers some resistance to the flow of electric current, and that resistance produces heat. The amount of resistance and consequent amount of heat produced are determined by the material from which the conductor is made, the size of the conductor, and the amount of current flowing. Therefore, if a rubber insulated wire, for example, is carrying its rated current, some heat will be produced, but no more than can be safely dissipated. However, if the current load is increased sufficiently, the heat is likely to ignite the insulation on the wire.

When an electric circuit is broken or the bare conductors of a circuit are brought close together, the current jumps the gap, producing an arc. The arc or flash produced is exceedingly hot, reaching around 6,000 degrees Fahrenheit (3,316 degrees Celsius). So efficient is the electric arc that it is used as the ignition source in internal combustion engines. It is apparent that, in the presence of gaseous fuel, a fire can be started easily by an electric arc, and any opening or closing of a circuit is apt to produce one. Loose connections, allowing even momentary opening and closing of a circuit, or the breaking of a light bulb may produce an arc that furnishes the ignition temperature for a union of fuel and oxygen.

Friction of one material upon another or of two pieces of the same material produces heat in proportion to the amount of the friction. This applies especially to bearings in machinery. Lubricating oil may help to reduce the amount of friction, but if a sufficient amount of oil is not supplied to a bearing, enough heat may be generated to ignite surrounding combustible materials. Friction of metals may be so severe as to cause sparks of fused metal that may become causes of ignition if they fall on combustible material.

There is also the heat of chemical action. To be sure, combustion is a chemical action, but in this case, let us consider only the slow generation of heat due to chemical action that may become great enough to cause ignition of the materials concerned or of such materials as may be heated by conduction. One group of materials (unslaked lime and several other chemicals) are not combustible but, on contact with water, generate sufficient heat to ignite combustible materials. Another group of materials, such as white phosphorus and aluminum triethyl in the presence of air and sodium and potassium in the presence of water, has ignition temperatures below ordinary temperatures. Still another group includes certain oils, fats, and other materials subject to spontaneous heating ignition.

Heat Transfer

Heat is transferred from one material or object to another by one or more of three methods — conduction, radiation, or convection.

By conduction, heat from one body is transferred to another by actual contact. Thus, a steam pipe passing through wood construction transfers its heat to the wood at the point where they touch one another. The amount of heat transferred in this way depends on the conductivity of the material through which the heat is passing. Conductivity varies greatly from material to material. Wood and solid asbestos have about the same conductivity, while fibrous materials, such as cotton, wool, and mineral wool, have considerably less conductivity. Brickwork and concrete have a heat conductivity about four to six times greater than wood. Iron has a heat conductivity 300 times as great, and aluminum, about 1,200 times as great as wood.

The rate of heat transfer through any material depends upon the temperature difference between the points of entrance and departure. The amount of heat flow also depends, among other factors, upon the size of the heat path. Just as more water will flow through a large pipe than through a small one under the same pressure, so more heat will flow through a large path of heat conductance than through a small one, assuming that the temperature difference is the same. The factor of temperature in the flow of heat may be likened to the factor of pressure in the flow of water.

The transmission of heat cannot be prevented by any heat insulating material. In this respect the flow of heat is unlike the flow of water, which can be stopped by a solid barrier. Heat insulating materials have a low heat conductivity, and, although heat flows through them slowly, no amount of insulating material can actually stop the flow. This fact should be remembered in devising protection against heat producing devices or other sources of heat that might ignite nearby woodwork. Filling the space between the source of heat and the combustible material with insulation is not sufficient to prevent ignition no matter how thick the insulation. If the heat continues to flow through for a sufficient length of time and if it does not escape somewhere, it will eventually raise the temperature of the woodwork. For this reason there should always be an air space between the heat source and nearby combustibles or some way of carrying the heat away, rather than relying solely on heat insulating material to protect exposed woodwork.

Radiation is the transfer of heat from one body to another by heat rays traveling through the intervening space, much in the same manner as light is transferred by light rays. Thus, heat comes to us from the sun. The heat from the steam pipe mentioned in an earlier example can be transferred, to some extent, to the wood construction by radiation, even though there is space between the two. A stove heats a room partly by radiation.

Radiated heat is not absorbed by the air to any great extent. Heat travels through space until it encounters a solid object, where it is absorbed and conducted through the object. Radiated heat is reflected by bright surfaces and will pass through glass. Heat radiation is a two-way process. Heat radiates from the stove to the wall; the wall, in turn, radiates heat in all directions when heated above the temperature of other objects in the room. The effect of heat is practically independent of distance, except that the rays from a heat source spread out in all directions. Therefore, the farther you are from the source, the less is the concentration that will reach you.

Convection is the transfer of heat by a circulating medium — either a gas or a liquid. Heat generated in a stove raises the temperature of the air surrounding it by conduction. The heated air circulates throughout the room and heats distant objects by convection. Since heated air expands and rises, convection is mostly in an upward direction, though air currents can carry heat by convection in any direction. The ceiling above a stove is heated by radiation and convection. The floor under the stove is heated only by radiation from the bottom of the stove, unless it is a type without legs that rests on the floor, in which case heat is conducted downward through the metal of the stove.

Hazards of Combustion

For his own safety, the fire fighter should know something about the nature of combustion. Ordinary combustible materials, such as wood (composed principally of carbon and hydrogen), produce carbon dioxide and water (steam) when burned completely, both of which are relatively harmless unless they deplete the normal oxygen supply in the air or raise the temperature of the air so high as to be unsafe for breathing. However, when carbon or materials containing carbon burn in a restricted air supply, carbon monoxide is formed. This is a poisonous and explosive gas — the well-known poisonous constituent of automobile exhaust gases.

There are other more or less dangerous gases that may be produced in fires. Most combustible materials, such as wood, are broken down by heat to form combustible gases. The flame is the burning gas, and for the most part, the actual combustion is that of the gases produced rather than the wood itself. Gases are often produced in fires, under conditions where they do not burn completely but accumulate in the building and flash or explode later when there is an additional air supply. Various poisonous gases may be produced from the burning of leather, fur, woolen fabrics, and all sorts of other materials commonly found in buildings. Wherever chemicals are involved in fires, there is danger from the possible presence of toxic gases. Sulfur or compounds containing sulfur produce sulfur dioxide, which is a highly irritating and choking gas. Luckily this is so objectionable that there is little danger of anyone unknowingly breathing it in dangerous quantity as in the case of carbon monoxide.

Fire fighters who understand the nature of fire gases appreciate the necessity for ventilating burning buildings and the need for using breathing apparatus when working in buildings heavily charged with smoke. If proper precautions are taken, the personal hazard from fire gases can be minimized.

Explosions present another hazard for the fire fighter. There are several kinds of explosions, but the fire fighter is most likely to meet those caused by extremely rapid combustion, such as occurs when a flammable gas or dust is mixed with air and burns, all within a fraction of a second. (A back-draft or "hot-air" explosion, encountered in fire fighting, is this kind of explosion.) Another type of explosion occurs in dynamite and other explosives that suddenly decompose and liberate large quantities of gases when detonated. Still another kind of explosion may occur when a liquid or gas is heated in a closed container with a progressive increase in pressure, until the container breaks with sudden release of pressure. A boiler explosion is a typical example of this type, and such an explosion can occur in a closed tank of water (without a relief valve). A fourth kind of explosion is that of an atomic bomb due to nuclear reaction.

LOCATING AND DEALING WITH FIRE HAZARDS

A fire hazard is any condition favoring destruction of life or property by fire.

The principal groups of materials referred to as hazards would include:

(*a*) Light combustible materials — thin plywood, shingles, shavings, cotton and other fibers, paper.
(*b*) Combustible dusts.
(*c*) Flammable and combustible liquids.
(*d*) Flammable gases.
(*e*) Certain plastics.
(*f*) Combustible metals.
(*g*) Materials subject to spontaneous heating.
(*h*) Explosive materials, acids, oxidizing agents.

The principal sources of ignition referred to as hazards would include:

(*a*) Open flames and heaters — smoking, torches, gas flames, lamps, furnaces, ovens, heaters.
(*b*) Friction — hot bearings, rubbing belts, grinding, shredding, picking, polishing, cutting, drilling.
(*c*) Electricity — arcs and sparks, including lightning and static, heated resistances.
(*d*) Chemical reactions, particularly those liberating heat.

Conditions favorable to fire may be inherent to the building or to the processes and materials of the occupancy of the building. A fire hazard exists if any combustible material, the burning of which would cause loss of property and jeopardize life, is exposed to a source of ignition. A barrel of gasoline may be just as safe as a barrel of water, so long as the gasoline is confined in the barrel and is not subject to a dangerous degree of heat that might cause the container to fail. But if it is spilled or otherwise allowed to vaporize and mix with the air in the proper proportions, a hazard exists because the conditions favor a fire. Electric wiring is not a hazard unless the condition of the wiring, due to poor workmanship, deterioration, or current overload, is likely to produce a fire. A wood frame building may house an occupancy that presents no fire hazards and be considered safe; but if an adjoining building burns, the frame building becomes a fire hazard to its own occupancy. Likewise, a heating plant becomes a fire hazard only when building and occupancy conditions are such that the plant is liable to become a cause of fire.

For a fire hazard to exist, there must be a condition such that combustible material can be exposed to a possible ignition source. If it were not for sparks and cigarettes and other sources of ignition, gasoline vapors around a filling station would be no great hazard.

Housekeeping Hazards

One of the most important, and simplest to achieve, methods of minimizing the risk of fire occurring is good housekeeping. All members of the fire brigade should be familiar with the procedures and conditions that comprise good housekeeping. In plants where the fire brigade is responsible for fire prevention activities, the fire chief may be required to develop routines for, and assign personnel for monitoring and reporting on, housekeeping practices. Some tasks should be performed daily, and others, less frequently.

Following are some of the conditions and practices that should be checked regularly.

(*a*) Rubbish, waste, and other trash should be removed from the plant daily (more often if necessary) and stored safely until they can be disposed of.

(*b*) Special cans for the storage of oily waste or rags should be provided and used. Such cans have self-closing covers and are set up off the floor. If the contents do ignite while in such a can, no damage is likely to result.

(*c*) Metal, industrial waste barrels should be provided and used for the storage of ordinary waste and rubbish until it can be removed from the building. Such containers are designed to snuff out accidental fires involving their contents and to limit their external surface temperature.

(*d*) Often shallow metal pans and oil-absorbing compounds are used to control oil dripping from industrial machinery, such as lathes and milling machines. Oil-soaked material in these pans should be removed at regular intervals.

(The Protectoseal Co., Chicago)

Figure 2-1. Self-closing can for oily waste and wiping rags.

(Justrite Manufacturing Co., Chicago)

Figure 2-2. Self-closing safety cans for small quantities of flammable liquids.

(*e*) Hazardous materials should be handled in accordance with the safety rules of the plant. Flammable liquids, for example, should be kept in safety cans designed for that purpose.

(*f*) Combustible metal scraps should not be allowed to accumulate in large quantities. Those pieces destined to be salvaged or reclaimed should be kept in clean, dry, tightly covered, noncombustible containers and stored outdoors.

(*g*) Clean waste and packing materials should be stored in covered bins.

(*h*) Fire prevention personnel should be on the lookout for stock and equipment that may be placed so as to block fire doors, emergency exits, and access to sprinkler valves.

(*i*) In certain parts of a plant, it may be necessary, for reasons of safety, to ban smoking. These areas should be properly posted, and checks made to be sure the prohibition is being observed.

Heating Hazards

Industry employs a variety of heat-producing and heat-utilization devices, such as furnaces, forges, retorts, kettles, kilns, and others, which may be heated by gas, oil, solid fuel, or electricity. The factors discussed on the generation and transfer of heat have practical application in the prevention of fires caused by these devices.

Particularly important is the effect of continuous heat through the use of the equipment. During design and construction, attention is given to the fire box or burning chamber, interior lining and exterior covering, the arrangement of burners and grates, the means of draft, the removal of excessive heat and explosive gases, and the means of removing ashes and residue. In the installation of furnaces and heating devices, especially those producing high temperatures, it is obvious that ventilation, air spaces, and insulating material must receive careful consideration. However, these are matters for specialists in heating equipment. The fire brigade member should be concerned primarily with the maintenance of proper clearances and ventilation and with any obvious conditions that may contribute to the hazard. The brigade member should also be on the lookout for attempts to bypass or override safety devices.

Chemicals, Paints, and Oils

Various chemicals, paints, and oils are likely to be found in most manufacturing occupancies and in some mercantile properties.

Some chemicals are hazardous by themselves. Others are hazardous only when in contact with substances with which they combine very easily. Where chemicals of any kind are used, their nature, as well as their proper storage and handling, should be determined.

Certain noncombustible chemicals are oxidizing agents, which promote rapid combustion of other materials. They provide the oxygen for combustion without the need of an air supply. Under some conditions, the combustion may be so rapid that it becomes an explosion. Various nitrates, chlorates, and peroxides are dangerous when mixed with combustible material but are relatively harmless when stored by themselves in noncombustible buildings.

Stocks of white lead, zinc, linseed oil, and colors ground in oil do not present any special hazards within themselves. The oil, however, is combustible, and will burn freely if heated to its ignition temperature. Most paint thinners and dryers are classified as flammable or combustible liquids and should be handled accordingly. Usually, lacquers are flammable liquids. The handling and use of paints also present serious hazards when such materials as rags, paper, sawdust, and other combustibles become soaked and are left to ignite spontaneously. Spray painting is usually more hazardous than brush painting because of the volume

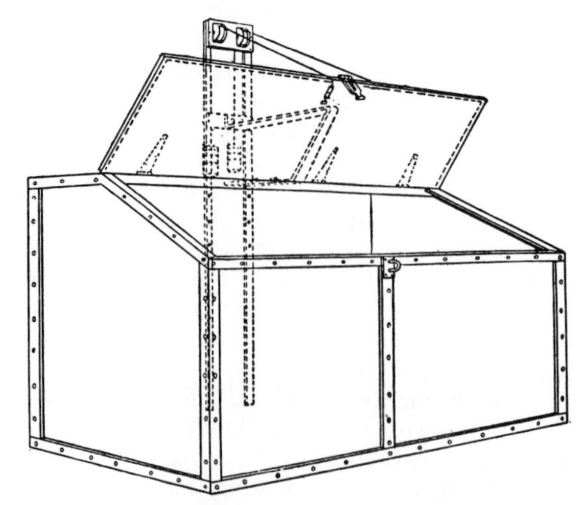

Figure 2-3. Metal bin with automatic cover for the storage of combustible packing material.

of flammable liquids used, the methods of application and drying, and the formation of combustible residue.

Spontaneous Heating

Spontaneous heating is an oxidation process in which the temperature of a material is increased without drawing heat from the material's surroundings. Heating of a substance to its ignition temperature in this manner results in spontaneous ignition. Whether the ignition temperature is reached depends, in part, on the rate of heat generation, the air supply, and the insulating properties of the immediate surroundings.

Just about all organic substances capable of combining with oxygen will oxidize at some critical temperature. With most materials, oxidation at normal air temperature is so slow that the heat evolved is dissipated to its surroundings as rapidly as it is formed; therefore, the temperature of the material does not increase. However, some combustible materials generate heat more rapidly than it can be dissipated; spontaneous ignition is the result. In order for spontaneous heating to take place, sufficient air must be available to support oxidation, but not so much that heat is carried away as quickly as it is generated.

Some vegetable oils, especially linseed oil, oxidize at normal air temperature. When spread out as in paint, the oils oxidize to a hard, tough film. Oxidation of such oils produces more than the ordinary amount of heat; however, it is quickly dissipated. If linseed oil or paint having linseed oil as a vehicle is wiped up with a rag and the rag is loosely wadded, oxidation with accompanying heat generation occurs. If the heat produced does not escape to the air, it will build until the ignition temperature of the oily mass is reached. If the rag is hung in the open air, the heat of oxidation will be carried away by air circulation. Conversely, if the rag is in a tightly packed bale, there may be insufficient oxygen within the bale to support oxidation.

Animals oils that become rancid oxidize and produce heat. Mineral oils at ordinary temperatures have little affinity for oxygen; therefore, they are less hazardous than vegetable oils.

The moisture content of hay and alfalfa has a definite effect on spontaneous heating in those products. Improperly cured before storage, they are very likely to undergo spontaneous heating. Meal, seeds, and beans having a high content of oxidizable oils are also subject to spontaneous heating.

Dust Explosions

The dust from any ordinary combustible material or oxidizable metals will, when mixed with air in the proper proportions, explode if ignited by a flame or spark. Any dust and air mixture is not necessarily explosive, because the proportion of dust to air necessary for ignition is quite high; consequently, dust explosions are not so frequent as might be expected at first thought. However, when they occur, they are quite disastrous.

Generally, two or more distinct explosions occur; the first or primary explosion may be in a comparatively small area that is in direct contact with the ignition cause. This primary explosion jars accumulated dust loose, throwing it into the air, and a secondary and quite extensive explosion may result. The explosion may follow open chutes, stairways, or elevator shafts.

The explosion hazard of a particular dust depends not only upon the combustibility of the material from which it is produced, but also upon the fineness and condition of the dust.

Dust explosions can be prevented in one or more of three ways — by eliminating dust production, by collecting and disposing of dust as it accumulates, and by eliminating open flames or other possible sources of ignition.

Dust may be ignited by flame or a spark. The sources of sparks are a matter not fully determined, but to realize the dangers involved, one has only to think of the possibility of sparks around moving machine parts and sparks from tramp metal in combustible material being ground. Static electricity produced by moving belts and conveyors may be a cause of dust ignition.

Flammable Liquids

The distinction between a flammable liquid and a liquid that is combustible but not flammable lies in the ease with which it gives off flammable vapors. For example, gasoline, alcohol, and acetone are flammable; lubricating, vegetable, or fish oils, or glycerine are combustible but not flammable. When any combustible liquid is heated to its flash point, it has about the same fire hazard properties as a flammable liquid. It may also become dangerously flammable if dispersed in a very fine spray.

The lowest temperature at which a liquid gives off vapors that can be ignited by a spark or flame in a test apparatus is called flash point of the liquid. The ignition temperature is usually much higher. Do not confuse flash point and ignition temperature. A flammable liquid can be above its flash point and not ignite unless there is a spark or flame — for example, gasoline at room temperature in an open pail.

Flash point is measured in two ways — closed cup in a standard closed apparatus and open cup in an open apparatus. Open cup flash point figures are higher. Flammable liquids, as defined in NFPA 30, *Flammable and Combustible Liquids Code*, have closed cup flash points below 100° Fahrenheit (37° Celsius). Liquids with flash points at or above 100° Fahrenheit

(37° Celsius) are combustible liquids. The lower the flash point, the greater the hazard. Liquids having a flash point lower than the temperature of the room where they are used are particularly hazardous because they give off flammable vapors whenever they are exposed to the air. Some examples and their closed cup flash points are carbon disulfide, −22°F (−30°C); ethyl ether, −49°F (−45°C); gasoline, −45°F (−42°C); and ethyl alcohol, 55°F (12°C).

The flammable range of a flammable vapor mixed with air lies between the lower flammable limit, which is the percentage of vapor below which a fire or explosion cannot occur (mixture too lean), and the upper flammable limit, which is the percentage of vapor above which there is not sufficient air for a fire or explosion to occur (too rich).

Flammable liquid vapors are heavier than air and, therefore, may accumulate in basements and pits if not dispersed by ventilation. Vapors can flow unseen, finally reaching some source of ignition and flashing back.

Identification of flammable liquids is often difficult. Smell is not a reliable indication, but it may be a signal that investigation of the hazard is needed. Trade names usually give no clue to the fire hazard. In case of doubt, get a laboratory test of flash point.

The hazard depends to a certain extent upon the relative quantity of flammable liquid exposed and the volume of the room or building where it is used. An increase in quantity of flammable liquid may not increase the probability of ignition, but would provide more fuel for continuing a fire.

Closed containers should be used as far as possible for storage and handling of flammable liquids. Pumps are the best means of withdrawing a flammable liquid from tanks or drums. Sources of ignition should be excluded from any room where flammable liquids are used or where there is any likelihood of leaks of flammable liquids from piping or closed containers. No smoking, no open flame heating or process equipment, and use of explosion-proof electrical equipment are common precautionary measures.

Fires in flammable liquids spread so rapidly that the entire exposed surface may be involved before any extinguishing agent can be applied. The extinguishing equipment provided must be sufficient to deal with a fire involving the entire liquid surface. This is unlike ordinary combustible materials where a relatively small extinguisher can be used on a fire before it has had time to spread.

Fires in flammable liquids generate so much heat and smoke that it is usually impossible to remain in a room with a flammable liquid fire for manual fire fighting unless the fire is small in relation to the size of the room. Do not be misled on this point by fire extinguishing demonstrations outdoors. Automatic extinguishing equipment is most desirable for operations where any quantity of flammable liquid is exposed indoors, as in an open tank.

The hotter a flammable liquid, the more rapidly vapors are given off. Extinguishment is more difficult after a fire has been burning long enough to heat the liquid and surroundings. After extinguishment, reignition is likely if glowing embers or hot metal surfaces are not cooled.

IDENTIFYING HAZARDOUS MATERIALS

For their own safety, fire brigade members should know where hazardous materials are located and the nature of the hazard. The NFPA has developed a labeling system for use on fixed installations that indicates the hazards to be found in the area.

Rail or highway accidents involving vehicles carrying hazardous materials may endanger the communities in which they occur. Therefore, the United States Department of Transportation (DOT) requires placarding of railway cars, trucks, and trailers carrying hazardous materials, so that fire and police personnel will be alerted to the danger. Since such vehicles will be on industrial plant property from time to time, fire brigade members should know the DOT system as well as the NFPA system. Both are discussed here.

NFPA System

A recommended system provides simple, readily recognizable and easily understood markings that will give at a glance a general idea of the inherent hazards of any material and the order of severity of these hazards as they relate to fire prevention, exposure, and control.

This system is explained in the table on page 13. It identifies the hazards of a material in three categories — health, flammability, and reactivity — and indicates the order of severity in each of these categories by five divisions ranging from "4," indicating a severe hazard, to "0," indicating no special hazard.

While this system is basically simple in application, the hazard evaluation that is required for the precise use of the signals in a specific location must be made by experienced, technically competent persons. Their judgment must be based on factors encompassing a knowledge of the inherent hazards of different materials, including the extent of change in behavior to be anticipated under conditions of fire exposure and control.

In general, health hazard in fire fighting is that of a single exposure which may vary from a few seconds up

IDENTIFICATION OF FIRE HAZARDS

HEALTH	FLAMMABILITY	REACTIVITY
Possible Injury	**Susceptibility of Materials to Burning**	**Susceptibility to Release of Energy**
Color Code: BLUE	**Color Code: RED**	**Color Code: YELLOW**
4 A few whiffs of the vapor could cause death or the vapor or liquid could be fatal on penetrating the fire fighter's normal full protective clothing which is designed for resistance to heat. The normal protective clothing will not provide adequate protection against skin contact with these materials.	**4** Very flammable gases, very volatile flammable liquids, and materials that in the form of dusts or mists readily form explosive mixtures when dispersed in air. Shut off flow of gas or liquid and keep cooling water streams on exposed tanks or containers. Use water spray carefully in the vicinity of dusts so as not to create dust clouds.	**4** Materials which are readily capable of detonation or of explosive decomposition or explosive reaction at normal temperatures and pressures. Includes materials which are sensitive to mechanical or localized thermal shock. If a chemical with this hazard rating is in an advanced or massive fire, the area should be evacuated.
3 Materials extremely hazardous to health, but areas may be entered with extreme care. Full protective clothing, including self-contained breathing apparatus, rubber gloves, boots and bands around legs, arms and waist should be provided. No skin surface should be exposed.	**3** Liquids which can be ignited under almost all normal temperature conditions. Water may be ineffective on these liquids because of their low flash points. Solids which form coarse dusts, solids in shredded or fibrous form that create flash fires, solids that burn rapidly, usually because they contain their own oxygen, and any material that ignites spontaneously at normal temperature in air.	**3** Materials which are capable of detonation or of explosive decomposition or of explosive reaction but which require a strong initiating source or which must be heated under confinement before initiation. Includes materials which are sensitive to thermal or mechanical shock at elevated temperatures and pressures or which react explosively with water without requiring heat or confinement. Fire fighting should be done from a protected location.
2 Materials hazardous to health, but areas may be entered freely with self-contained breathing apparatus.	**2** Liquids which must be moderately heated before ignition will occur and solids that readily give off flammable vapors. Water spray may be used to extinguish the fire because the material can be cooled to below its flash point.	**2** Materials which are normally unstable and readily undergo violent chemical change but do not detonate. Includes those materials which may react violently with water or which may form potentially explosive mixtures with water. In advanced or massive fires, fire fighting should be done from a protected location.
1 Materials only slightly hazardous to health. It may be desirable to wear self-contained breathing apparatus.	**1** Materials that must be preheated before ignition can occur. Water may cause frothing of liquids with this flammability rating. Most combustible solids have a flammability rating of 1.	**1** Materials which are normally stable but which may become unstable at elevated temperatures and pressures or which may react with water with some release of energy but not violently. Caution must be used in approaching the fire and applying water.
0 Materials which on exposure under fire conditions would offer no health hazard beyond that of ordinary combustible material.	**0** Materials that will not burn.	**0** Materials which are normally stable even under fire exposure conditions and which are not reactive with water.

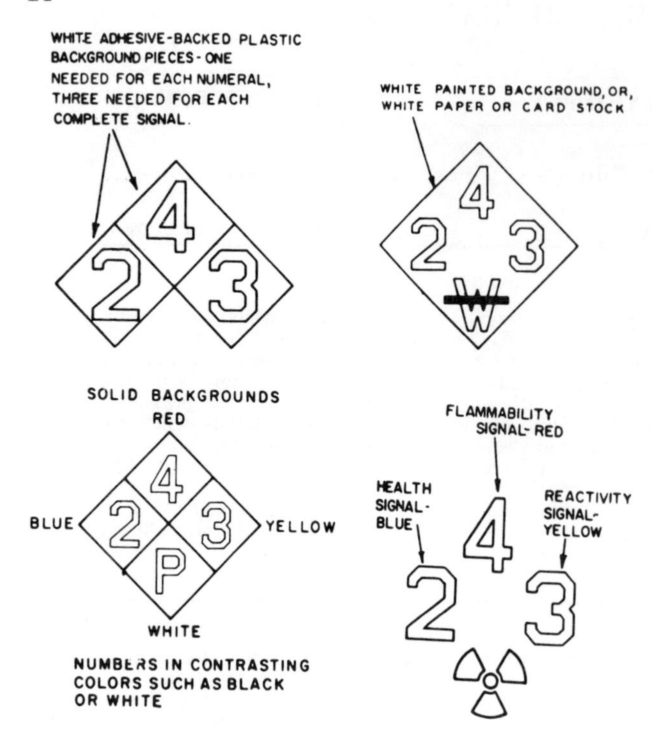

Figure 2-4. Methods of presenting NFPA 704 System hazard information.

Figure 2-5. Examples of two DOT placards. Various symbols, letters, colors, and combinations of colors are used to denote the nature of the hazard.

to an hour. The physical exertion demanded in fire fighting or other emergency conditions may be expected to intensify the effects of any exposure. Only hazards arising out of an inherent property of the material are considered. The protective equipment normally used by fire fighters is taken into account.

Susceptibility to burning is the basis for assigning degrees within the category of flammability. The method of attacking the fire is influenced by this susceptibility factor.

The assignment of degrees in the reactivity category is based on the susceptibility of materials to release energy either by themselves or in combination with water. Fire exposure is one of the factors considered along with conditions of shock and pressure.

A fourth space in the identification symbol is reserved for additional information that may be of value to the fire fighter. For example, any material that will react violently with water is noted with the symbol "W" with a line drawn through it to indicate "do not use water." Radioactivity can be identified in this space as well as other warnings affecting the extinguishing method to be employed.

DOT Placarding System

Trucks and trailers carrying certain hazardous materials must be marked front and rear and on both sides with the name of the hazard on diamond shaped placards specified by the Department of Transportation. Class A or B explosives, extremely toxic materials, and high strength radioactive materials in any amount must be identified.

Highly toxic materials, Class C explosives, flammable and combustible liquids, flammable and nonflammable gases, oxidizers, corrosives, irritants, organic peroxides, and certain poisons must also be identified when they are present in quantities of 1,000 pounds or more. If the cargo contains two or more materials in this group, the vehicle need only be placarded DANGEROUS. However, if the mixed cargo contains 5,000 pounds or more of one class of material loaded at a single facility, the placard specified for that class of material must be applied to the vehicle.

The placarding requirements for rail cars are essentially the same as those for motor vehicles. The same identifying placards are used. However, rail car placards for Class A explosives, poison gas, and POISON GAS — EMPTY must be placed over a square black-bordered white background.

An empty tank car that had contained a hazardous substance must carry a placard indicating the hazard of its last contents and the word EMPTY. If the tank car has been reloaded with a material not requiring placarding or has been cleansed of residue and purged of vapor sufficiently to remove the hazard, no placard is required.

Suggested Reading

Fire Protection Handbook, 14th Edition, 1976, NFPA, Boston, Sections 2, 3, 4, and 5.

NFPA Inspection Manual, 4th Edition, 1976, NFPA, Boston.

Fire Protection Guide on Hazardous Materials, 6th Edition, 1975, NFPA, Boston.

Chem-Cards (a series of cards giving information on hazardous chemicals), Manufacturing Chemists' Association, Washington, DC.

Chapter 3

FIRE EXTINGUISHERS

Most fires are relatively small when they start, and, if they are detected quickly, they can be easily handled with portable fire extinguishers. Therefore, industrial fire fighters should be familiar with the types and locations of fire extinguishers distributed throughout the plant. Brigade members should know how fire extinguishers work, how they are used, and how they are maintained.

Three methods of extinguishing fire have long been recognized — cooling the fuel below its ignition temperature, diluting the supply of oxygen contributing to combustion, and removing the fuel. There is a fourth method that applies only to flaming combustion. It involves chemical inhibition of the flame by the extinguishing agent. The exact mechanism by which flame inhibition takes place is not fully understood and is still a subject of study.

To cool a fuel, the extinguishing agent must have the ability to absorb enough heat to lower the fuel temperature below the ignition point. Water, which has very good heat absorbing qualities, is commonly used for this purpose. As water is played directly on the fire, heat is transferred from the fuel to the water by conduction. Eventually, the water is converted to steam, which removes heat from the fuel surface by convection.

Extinction by oxygen dilution is accomplished by blanketing the fuel surface with a noncombustible gas. The gas displaces, and mixes with, the surrounding air. As increasing amounts of gas are introduced, the percentage of oxygen in the mixture decreases. When the oxygen level falls below that which will support combustion, the fire goes out. Carbon dioxide is commonly used to achieve fire extinguishment by oxygen dilution. The generation of steam by the application of water to fires in enclosed spaces or compartments also contributes to oxygen dilution.

Fire may also be extinguished by removing or cutting off the fuel. Pumping a flammable liquid from a storage tank that is involved in fire to one that is not is one form of fuel removal. Another is the removal of stored raw materials or commodities that are in the path of a slow burning fire. Examples of cutting off the fuel are closing a valve when the fuel is escaping gas, blanketing a solid or liquid fuel with fire fighting foam, or covering combustible metal fires with an inert material, such as sand.

Combustion is a chain reaction, and the fourth method of extinguishment uses agents that break the chain. Certain halogenated hydrocarbons, alkali metal salts, and ammonium salts react chemically with flames to extinguish fire. Typical halogenated agents are bromotrifluoromethane (Halon 1301) and bromochlorodifluoromethane (Halon 1211). Alkali metal salts used to extinguish fire by flame inhibition are sodium bicarbonate, simply called dry chemical; potassium bicarbonate, whose trade name is Purple K; potassium carbamate, whose trade name is Monnex; and potassium chloride, whose trade name is Super K.

GENERAL PROCEDURE IN USING FIRE EXTINGUISHERS

Portable fire extinguishers are intended to be used against small fires; therefore, they are designed to be used close to the burning material. It is important for the operator to learn the capabilities and limitations of the various types of appliances.

Fire extinguishers should not be expected to take the place of automatic sprinklers and hose streams or of specially designed extinguishing systems for controlling larger fires. For example, consider the case of a fire involving a flammable liquid in an open tank. Considerable heat and smoke would be given off. Portable fire extinguishers can handle a fire in such a tank. However, if the tank is inside a building, the heat and smoke buildup may be so great and so rapid as to prevent a person from getting near the fire. Usually, it is a question of how much flammable liquid surface is exposed. For effective use of extinguishers, the maximum surface area for such tanks has been found to be about 30 square feet (2.8 square meters), but appropriately engineered extinguishing systems should be considered when surface areas exceed 20 square feet rather than depending on the use of portable extinguishers. Automatic extinguishing systems are recommended for certain dip tanks having liquid surface areas exceeding 4 square feet (0.4 square meter).

No fire should be attacked with portable extinguishers alone. A good rule is to send someone to call the plant fire brigade or the public fire department. If you are alone and a fire occurs, summon assistance first and then try to put out the fire with an extinguisher. This is not to discourage the use of fire extinguishers, but to emphasize that a person using them should know what the extinguisher is designed to do and what can be expected of it. Furthermore, much will depend upon the skill with which the extinguisher is used. Therefore, industrial fire fighters should seize every opportunity to familiarize themselves with the particular units provided in the plant and practice using them as often as possible.

TYPES OF EXTINGUISHERS

Present-day Types

Portable fire extinguishers currently being manufactured use one of five principal types of extinguishing agent — water, compressed gas, dry chemical, dry powder, or liquefied gas.

Water-filled extinguishers use stored pressure or pumping action to expel the extinguishant.

Compressed gas extinguishers utilize carbon dioxide both as the extinguishant and as a self-expellent.

Dry chemical extinguishers contain such agents as specially treated sodium bicarbonate, potassium bicarbonate, potassium chloride, or ammonium phosphate. The agent is expelled from the appliance either by stored pressure or by pressure supplied by a small cartridge of carbon dioxide or nitrogen.

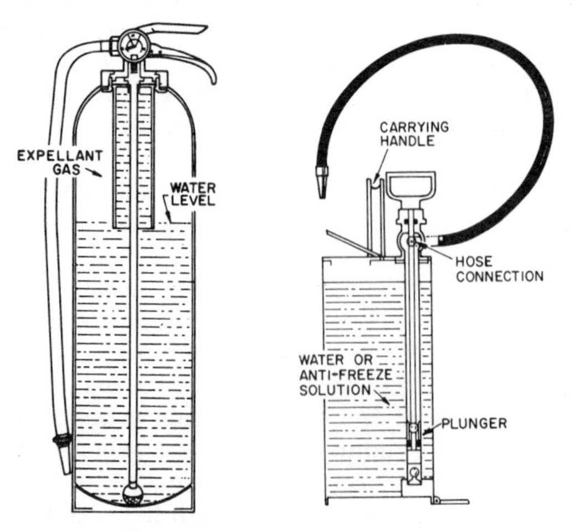

Figure 3-1. Water-filled extinguishers — stored pressure (left) and pump tank (right).

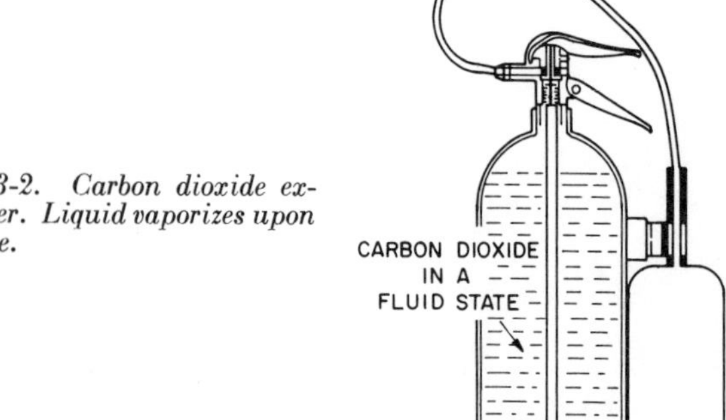

Figure 3-2. Carbon dioxide extinguisher. Liquid vaporizes upon discharge.

Dry powder extinguishers are cartridge-operated and contain sodium chloride as the extinguishing agent.

Liquefied gas extinguishers contain either bromotrifluoromethane (Halon 1301) or bromochlorodifluoromethane (Halon 1211). The agent is a self-expellent, but usually stored pressure (nitrogen) is used to improve discharge characteristics.

Obsolete Types

The manufacture of the invert-to-operate type of fire extinguisher was discontinued in the United States in 1969. Included in this category are soda-acid, foam, and cartridge-operated water and loaded stream extinguishers. After ten or fifteen years of service, fire extinguishers of these types displayed an alarmingly high failure rate when pressurized. This as well as the inconvenient method of operation helped to bring about the discontinuance of these extinguishers. Though they are no longer tested and listed by Underwriters Laboratories Inc., many of these extinguishers are still in use. Therefore, they are included in the discussions in this chapter.

Earlier in the 1960's, vaporizing liquid fire extinguishers using carbon tetrachloride and chlorobromomethane were discontinued because the toxic properties of these agents posed a serious health hazard to persons operating the extinguishers. Fire extinguishers containing those agents should not be used at all and are not discussed here.

SUITABILITY OF DIFFERENT TYPES

The suitability of a particular fire extinguisher is determined by the conditions under which it may be used. The variety of situations for which a portable

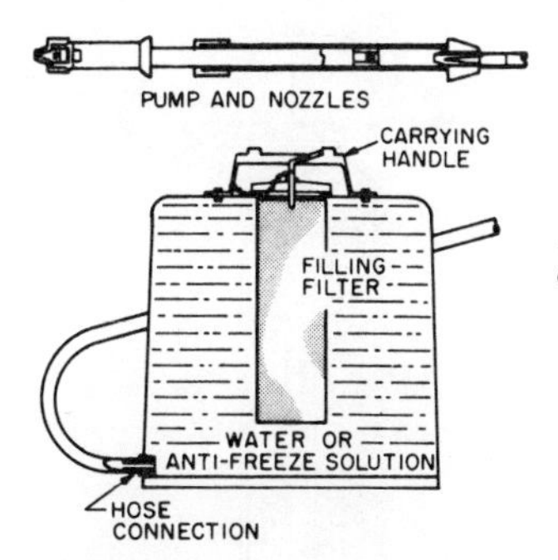

Figure 3-3. Backpack pump tank extinguisher.

fire fighting device is desirable has led to the development of a number of types and sizes of appliances. Water is the most efficient extinguishing agent for fires involving ordinary combustibles. Since more fires involve ordinary combustibles than involve other fuels requiring special extinguishing agents, water solution types of fire extinguishers are the ones most often provided. Nevertheless, there are situations where other agents are more effective than water or where the use of water solutions may introduce a hazard to the operator of the extinguisher.

Ordinary Combustibles

The majority of fires involve ordinary combustibles, such as paper, wood, and cellulosic (natural) fibers. For these materials, water solution extinguishers are the first choice. They will put out more fire in ordinary combustibles than will any of the other types, and the wetting effect of water prevents reignition.

Multi-purpose dry chemical extinguishers (ammonium phosphate base) are also suitable for fires involving ordinary combustibles. Following application, the agent forms a coating that swells and tends to retard further burning.

Carbon dioxide can, of course, be used on small fires in ordinary combustibles, but it is not a practical

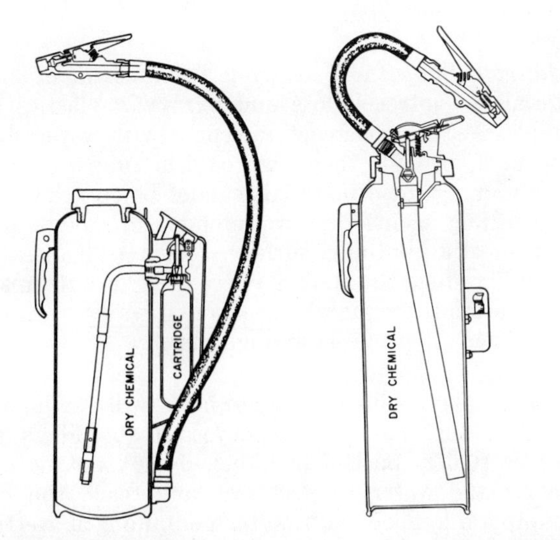

Figure 3-4. Dry chemical extinguishers — cartridge-operated (left) and stored pressure (right).

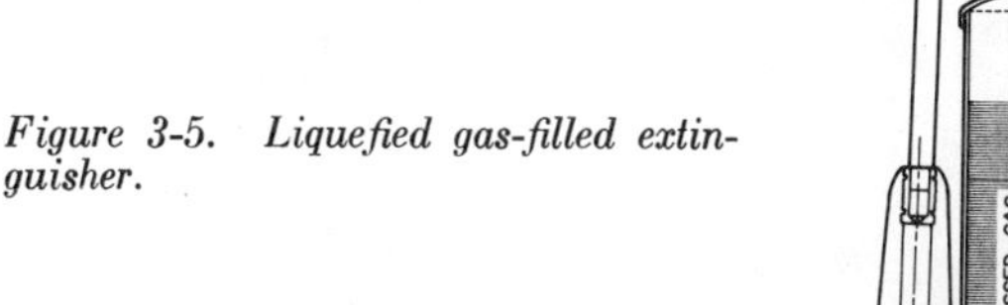

Figure 3-5. Liquefied gas-filled extinguisher.

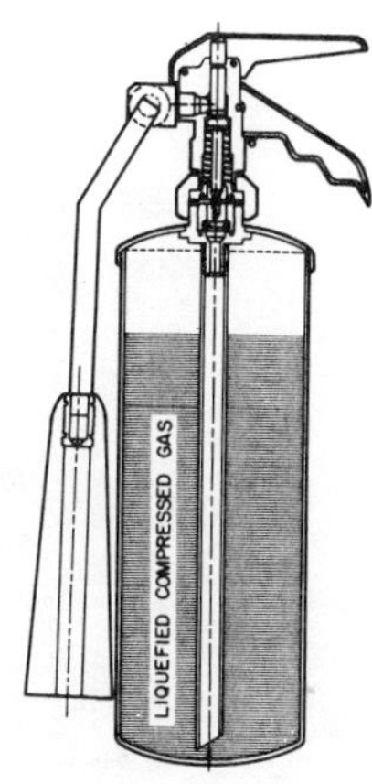

agent. Relatively large amounts of carbon dioxide gas would be required to extinguish a fire that could be extinguished by small amounts of water or multipurpose dry chemical, and rekindling is likely.

Bromochlorodifluoromethane (Halon 1211) extinguishers of 5½- to 22-pound capacity are rated for small fires involving ordinary combustibles.

The foam extinguisher can be used on ordinary combustibles and is effective to the extent that there is water in its solution. It does not penetrate as well as water, for example, into kindling or loose fibers.

Flammable Liquids

Fires in flammable liquids are often in open tanks, like tanks of paint solution or hardening and tempering oils. Fires also occur in spills of small amounts of liquid on the ground or floor. As has been pointed out, the amount of flammable liquid burning must not be so great that an operator with an extinguisher cannot get near the fire or cannot stay near enough to it to use the relatively limited capacity of portable extinguishers. The foam and carbon dioxide types are suitable for flammable liquids because they provide a smothering blanket over the burning surface of liquid exposed. Dry chemical and Halon types are effective because they have a chemical reaction that tends to interrupt the combustion process.

Extinguishers using carbon dioxide, dry chemical, or Halon are particularly suitable where a small amount of extinguishing agent can be promptly used to snuff out a fire. For fires in open tanks or involving large spills, the foam extinguisher may offer advantages by furnishing a persistent blanket over the surface, preventing reignition of the vapors. The chemical in the loaded stream type is often effective on flammable liquids in pans or tanks, but is a second choice to foam for such situations.

Electrical Equipment

For fighting fires in live electrical equipment, an extinguishing agent that produces a nonconducting

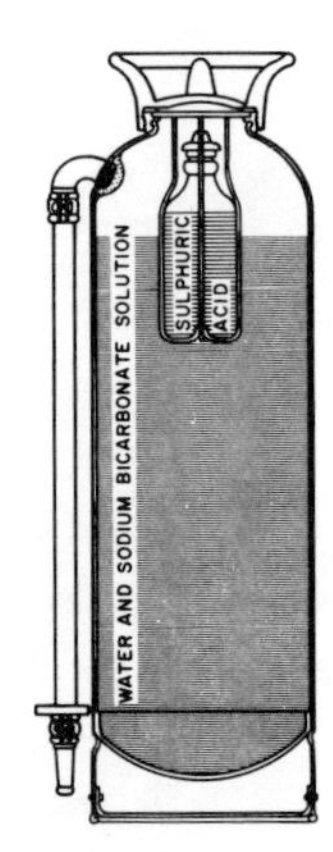

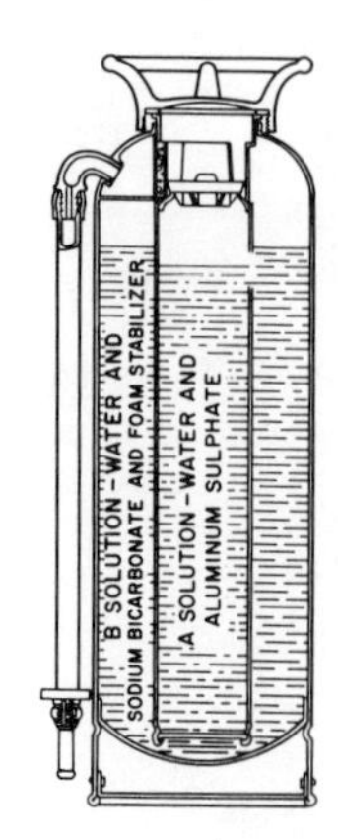

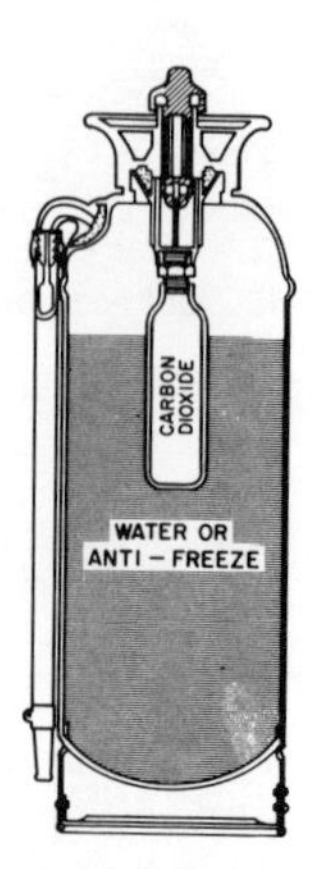

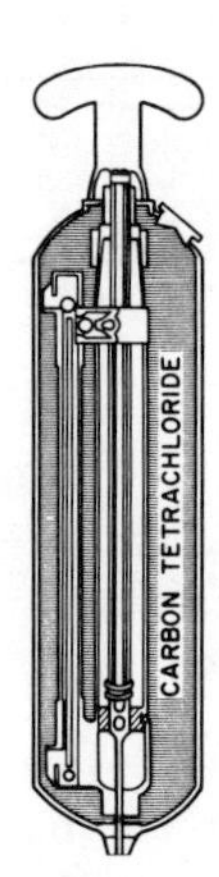

Figure 3-6. Obsolete portable fire extinguishers — (from left to right) soda-acid, foam, cartridge-operated water extinguisher, and vaporizing liquid extinguisher.

stream or discharge is of first importance. Carbon dioxide extinguishers can be used for most of these situations, usually involving electric motors, switch gear, and delicate electrical apparatus. Dry chemical units are also safe to use. For some equipment, dry chemicals may make the business of cleaning up and restoring operation of the equipment more difficult. Halon 1211, which is available in portable extinguishers, is also recommended for fires involving electrical equipment. Since this agent leaves no residue, the task of cleaning up is greatly simplified. The use of nonconducting discharge "horns" on fire extinguishers intended for use against fires involving live, exposed, electrical equipment is important. Where possible, electrical equipment should be deenergized (Chapter 7).

Combustible Metals

Carbon dioxide, most dry chemical, and liquefied gas extinguishers are ineffective on fires in which certain combustible metals are burning. Water solutions are ineffective unless the amount of burning metal is very small and the amount of water used is relatively large, so that the fire can be suddenly quenched. A small amount of water will be turned into steam with explosive force upon contacting the hot metal. A number of powders and liquids are provided for fires involving magnesium, titanium, zirconium, sodium, and potassium. On certain combustible metals (e.g., magnesium), coarse water streams from hose lines or sprinklers can control such fires after some initial temporary intensification, but portable water solution extinguishers cannot be used safely. Dry sand can be used in some cases to cover a small mass of burning metal, but this action usually does not extinguish the fire.

Special Situations

A considerable variety of special situations arise where extinguishers of one type or another should not be used or where certain types can be used, provided the operator thoroughly understands the uses and limitations of the equipment.

Fire in Confined Spaces: The use of certain types of fire extinguishers is limited by the size of the room or space in which the fire occurs. For example, if carbon dioxide is discharged in a small room, the added carbon dioxide gas in effect decreases the percentage of oxygen in the room, and it is conceivable that the operator could suffocate. Dry chemical dispensed in small enclosed spaces may produce a cloud of chemical, which could cause choking and restrict visibility. Liquefied gas extinguishers produce offensive gases. None of these types should be provided where an uninstructed operator might attempt to use them in too confined a space.

Oxidizing Agents: Many oxidizing agents — potassium nitrate is one—react violently with water solutions. As in the case with burning metals, it is not safe to use the small amount of extinguishing agent that is in a portable extinguisher on any but trifling quantities of these materials. Fires can be controlled if the burning material can be washed away with relatively large amounts of water. Inert gas extinguishers are ineffective because the burning materials themselves produce enough oxygen to maintain combustion even when smothered.

Materials Subject to Exothermic Decomposition: These materials — nitrocellulose and pyroxylin plastics are examples — do not react violently with water but, in burning, provide their own oxygen supply. Thus, fires involving these materials cannot be controlled by extinguishing agents that are intended to exclude oxygen from the burning surface. Water-filled extinguishers produce the cooling effect needed, but fires in these materials frequently need more water than can be provided in portable extinguishers.

Fibers: Water-filled extinguishers will extinguish fires in natural fibers. However, if the fibers are baled or tightly packed in piling, it may not be possible for the water to penetrate and reach the fire. A pump tank filled with water containing a wetting agent may be a suitable extinguisher for some cases of this sort.

Combustible Dusts: In a room where there are dust accumulations, as on shelves or beams, the jet of liquid or gas discharged from an extinguisher may dislodge a small cloud of dust that may ignite with a puff. The ignition may be violent enough to stir up larger clouds of dust and a major dust explosion could result. Grain dust is a common example of combustible organic dusts. Dusts of metals, an example of which is the fairly common aluminum power, are combustible and make a violent explosion. Water streams should be applied in a spray pattern where there is a chance that a solid stream may agitate combustible dust.

Methods of Operation

The method of putting each extinguisher into use should be studied and practiced fully. The common methods of operation are (a) to operate a lever, trigger, or wheel; (b) to pump; (c) to invert the unit; and (d) to invert and bump the unit on the ground.

Lever- or Wheel-Operated Types: In types where stored air pressure or pressure from compressed carbon dioxide, nitrogen, or other gas is used, the operation is started by pressing a trigger or lever or, less commonly, by turning a wheel. This opens a valve or cuts a disc to release the gas. Most modern extinguishers are provided with trigger grip operating controls located on the head of the device or at the nozzle.

Pump Types: The pump tank water units, as their name implies, are operated by pumping.

Invert Type: The traditional "chemical" extinguisher started the chemical reaction that produces the discharge pressure by turning the unit over. This action causes the chemicals to mix. The method is employed in the soda-acid and foam types.

Invert and Bump Type: In this type the discharge pressure is from a carbon dioxide cartridge. The bumping punctures a disc and releases the gas from the cartridge. Most of these units have the same general appearance as the conventional designs that are turned over, but the puncture of the disc is the essential operation. They are designed to be turned over partly to conform with tradition and partly because bumping is a convenient method of striking the blow to break the disc.

NUMBER AND LOCATION OF EXTINGUISHERS

The type, size, and number of extinguishers provided are usually matters of judgment. The kind of fire determines the type, and the size usually depends on who will use the units. The 5-gallon pump tank may be selected for locations requiring a water type extinguisher where personnel having sufficient strength and training are available to use it. For a hospital or office building, one of the 2½-gallon units, preferably of a type not requiring pumping, might be better because the users will more likely be persons without much training in the use of the devices and, in some cases, without strength to effectively handle the larger device.

The location of extinguishers also affects the total number provided. They should be close at hand in any case. The number of extinguishers needed to protect a property would be determined by the areas and the arrangement of the building or occupancy, the severity of the hazard, the anticipated classifications of fire, and the distances to be traveled to reach extinguishers. Other factors that need to be considered are the anticipated rate of fire spread, the intensity and rate of heat development, the smoke contributed by the burning materials, and the accessibility of a fire to close approach with portable extinguishers. It is obvious that each property has to be surveyed for actual protection requirements.

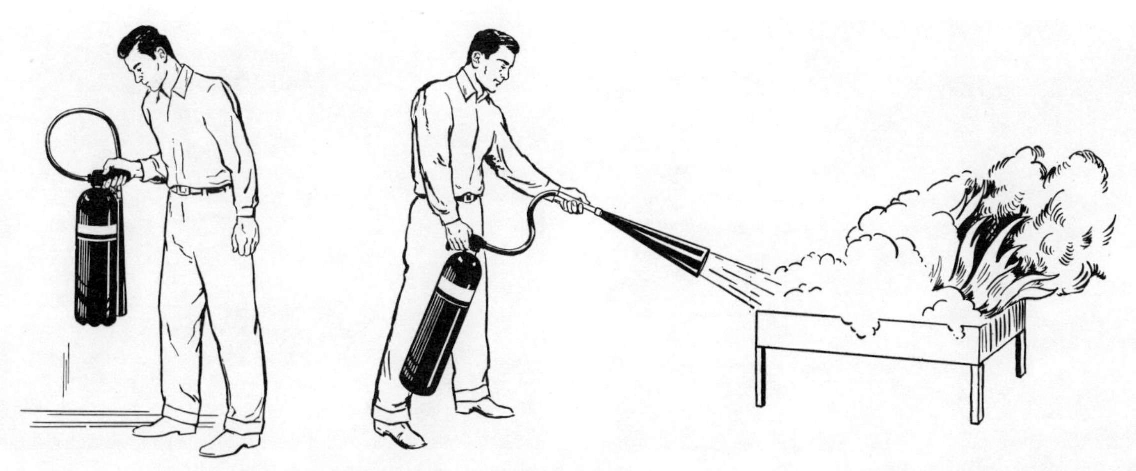

Figure 3-7. Using a carbon dioxide extinguisher. Lift the extinguisher off its bracket. At the fire, break the seal, pull the safety pin, and squeeze the trigger to start discharge. Direct the discharge as close to the fire as possible, applying it first at the edge and bottom of the fire and progressively moving it forward while sweeping the discharge from side to side. Continue discharge, even after the fire is out, to cool the liquid and prevent possible reignition.

Most buildings require Class A fire extinguishers, as ordinary combustible materials are common to virtually every occupancy. Rarely will a building require only Class A fire protection. A hospital, for example, will have great need for Class A extinguishers, covering such areas as the patients' rooms, corridors and offices, but will need Class B extinguishers in laboratories, kitchens and places where flammable anesthetics are stored or handled and Class C extinguishers for the protection of electrical switchgear or generator rooms.

Modern practice divides different occupancies into general classes of hazard termed "light," "ordinary," and "extra" hazard. A light hazard occupancy is one in which the amount of combustibles or flammable liquids present is such that fires of small size may be expected. Examples are homes, offices, schoolrooms, and churches. Ordinary hazard occupancies would include such areas as those used for mercantile storage and display, automobile showrooms, parking garages, light manufacturing and school shop areas. Where fires of severe magnitude may be expected, such as in woodworking plants, auto repair shops, and industrial plants involving processes in which flammable liquids are handled, the property is called extra hazard for the purpose of defining extinguisher needs.

Fire extinguishers for Class A fires should be so located that the maximum travel distance to extinguishers would not exceed 75 feet (22.8 meters) and, in general, the area protected per extinguisher should not exceed 3,000 square feet (279 square meters). For Class B and C hazards, the maximum travel distance should not exceed 50 feet (15 meters).

Where Class B fire extinguishers are being used to provide protection for such equipment as dip or quench tanks in which flammable liquids of appreciable depths are encountered, the extinguishers should be provided on the basis of one numerical unit of Class B extinguishing potential per square foot of flammable liquid surface of the largest tank within the area. When the liquid surface is in excess of 20 square feet (1.9 square meters), serious consideration should be given to installing fixed protection systems, and portable extinguishing protection for such hazards should generally be restricted to plants having trained fire brigades.

CARE AND MAINTENANCE

The management of a property should provide for the care and maintenance of portable fire extinguishers and related equipment. It may assign this work to its own maintenance force, or it may contract for the inspection and maintenance service. In some properties, the members of a fire brigade may be given some of the inspection and maintenance responsibilities.

Protection Against Freezing

The following extinguishers will not freeze at temperatures down to approximately minus 40 degrees Fahrenheit (minus 40 degrees Celsius).

- Water types including loaded stream and calcium chloride solution. The lowest safe temperature of the latter depends on the amount of calcium chloride in solution.
- Carbon dioxide.
- Dry chemical. This type may require nitrogen as a pressurizing medium.
- Liquefied gas.

Plain water, soda acid, and foam extinguishers must be protected from freezing — for example, by being installed in heated cabinets or by being kept at locations that are always above freezing or that are heated. Never try to change the solutions in soda acid or foam extinguishers with the idea of rendering them non-

(Ansul Chemical Co., Marinette, Wis.)

Figure 3-8. Using a 25-pound dry chemical extinguisher on a 125-square-foot pan fire. The operator first fans the discharging chemical rapidly back and forth across the near edge of the pan.

(Ansul Chemical Co., Marinette, Wis.)

Figure 3-9. The second step in the use of the dry chemical extinguisher on a pan fire is to push the fire back across the pan, finishing extinguishment at the far edge.

freezing. The action of these types depends on a particular chemical reaction, and any change in the recommended solutions is likely to render them inoperable. In the case of most plain water types, a calcium chloride nonfreezing solution may be used instead of plain water. Two and one-half gallons (9.5 liters) of such a solution with an approximate freezing temperature of minus 40 degrees Fahrenheit (minus 40 degrees Celsius) can be made with 10 pounds (4.5 kilograms) of flake 75 percent calcium chloride and 2 gallons (7.6 liters) of water.

Inspection

The inspection required is primarily to make sure that the extinguisher is in its designated place and that access to it is not obstructed. The inspection is also to determine that it has not been tampered with or used or that it is obviously inoperative.

Wire and lead seals, paper strips, and other indicators may be employed to help determine by a visual inspection whether the device has been moved or used. The units may be examined visually for dents and breakage and to see that caps, gages, valves, and hose are in place. Nozzles should be examined for clogging. Pressure gages should be tapped lightly to ensure that pointers are not stuck.

Inspections should be made at frequent intervals. Best practice is to have, for appropriate parts of the property, a list of the extinguishers. It should be required that the date of the inspection and initials of the person making the inspection be entered on such lists. The inspection should be more frequent than the record of recharging.

The relative importance of the extinguishers to the property will determine the frequency of inspection. Industrial and warehouse properties where access can be blocked by stock may require daily inspection. In most cases, extinguishers should be inspected monthly.

There are numerous factors that may determine the exact frequency of inspection. Among these are the property's experience with theft and tampering, the susceptibility of the units to leakage, mechanical injury, or damage by abnormal temperatures or corrosive atmospheres.

Maintenance

In addition to the daily or monthly inspection, at regular intervals as determined by conditions in the property, the extinguisher should be given a thorough examination. Certain types should always be discharged at this examination, which provides an opportunity for practice by the persons who may be called upon to operate the extinguishers.

In maintenance procedures, the mechanical parts of the unit should be examined. The extinguishing material should be examined to be sure that the proper

EXTINGUISHER MARKINGS

Water type extinguishers

Loaded stream extinguishers

Carbon dioxide and certain dry chemical extinguishers

Dry chemical multi-purpose extinguishers

Extinguishers and agents for metal fires

amount is available and in usable condition. The pumps provided in certain units should be operated. In stored pressure types, the examination should cover leakage and the amount of gas available.

The manufacturer's instructions should be explicitly followed in the periodic care and checking of the unit. The property owner who looks after his own extinguishers should use only the recharges furnished by the particular manufacturer. This procedure is also cheapest, as it would be prohibitively expensive for most users to have the technical staff and experience available to tell about the possible effects of substitutions on the successful operation of each type of extinguisher in use.

It is useful to keep a maintenance book or card record on each extinguisher in which maintenance work done periodically or after each use may be recorded. In addition, a durable tag should be attached to the extinguisher showing when it last had a thorough periodic examination or recharge.

Hydrostatic Testing

Numerous types of extinguishers employ pressure vessels. These types should be subjected to approved methods of hydrostatic tests at intervals specified for the particular type. (See NFPA 10, *Standard for Portable Fire Extinguishers.*)

WATER PAILS AND CASKS

Water pails are used for first-aid fire fighting but seldom except where there is some special local situation or where service on portable fire extinguishers may not be available. Water pails are less efficient than portable extinguishers. Their relative efficiency is usually judged to be five 12-quart (11.4-liter) pails or six 10-quart (9.5-liter) pails to achieve a rating comparable to the usual 2½-gallon (9.5-liter) water solution type of standard portable extinguisher: 2-A. A traditional water pail assembly is a covered water barrel or cask and three pails. Drums of the common 55-gallon (208-liter) capacity may be used in place of a barrel or cask. For places where space is at a premium, bucket-tanks of 25 to 55 gallons (94.6 to 208 liters) water capacity store the buckets or pails inside. Pails that are kept full should be provided with covers.

Suggested Reading

NFPA 10, *Standard for Portable Fire Extinguishers*, 1975, NFPA, Boston.

Fire Protection Equipment List, published annually, Underwriters Laboratories Inc., Chicago.

List of Equipment and Materials, Volume 1, Underwriters' Laboratories of Canada, Scarborough, Ontario.

Factory Mutual Approval Guide 1977, Factory Mutual Engineering Corporation, Norwood, Massachusetts.

TYPES OF FIRE EXTINGUISHERS

This list is limited to portable extinguishers of representative types and sizes. For special situations, other useful types and sizes are available as shown in manufacturers' catalogs and Underwriters Laboratories Inc. Fire Protection Equipment List. Underwriters Laboratories Inc. classifies extinguishers according to their fire extinguishing potential which is indicated by numeral and letter designations. The letter indicates the general class of fire for which the extinguisher is suitable following basic classes as defined in Standard for the Installation of Portable Fire Extinguishers, NFPA No. 10. Class A fires are those fires in wood, paper and the like. Class B fires are those in flammable liquids. (The number preceding a B-classification is based on tests of the device on fires which can be extinguished by a trained operator with the device). Class C ratings indicate only the electrical nonconducting characteristics of the agent. Agents for Class D fires, those in combustible metals, are not covered by the following table.

Description of Basic Types	Typical Nominal Capacity	Underwriters Laboratories Class	Weight Charged Pounds
Pump tank	2½ gallons	2-A	40
	4 gallons	3-A	
	5 gallons	4-A	65
Water (cartridge or stored pressure types and chemically-generated expellant operated)	2½ gallons	2-A	35
Antifreeze-calcium chloride solution (cartridge or stored pressure)	2½ gallons	2-A	40 to 45
Antifreeze-loaded-stream (cartridge or stored pressure)	2½ gallons	2-A:1-B to 3-A:1-B	40
Foam	2½ gallons	2-A:4-B to 2-A:6-B	70
Liquefied gas (bromotrifluoromethane)	2½ pounds	2-B:C	4
(bromochlorodifluoromethane)	5 pounds	——	—
Carbon dioxide	15 pounds	2-B:C to 10-B:C	50 to 60
Dry chemical (sodium bicarbonate base)	20 pounds	20-B:C to 40-B:C	35 to 45
Dry chemical (potassium bicarbonate base)	20 pounds	20-B:C to 60-B:C	35 to 45
Dry chemical (ammonium phosphate base)	2¼ to 2¾ pounds	5-B:C	40 to 60
Dry chemical (multi-purpose, ammonium phosphate base)	20 pounds	4-A:30-B:C to 10-A:40-B:C	35 to 45

Chapter 4

HANDLING HOSE

Fire hose must be able to withstand considerable abuse. It may be subjected to heat, mechanical abrasion, and corrosive materials. Yet through it all, fire hose must remain flexible, watertight, and durable. The industrial fire fighter should be aware of the types and uses of fire hose.

TYPES OF FIRE HOSE

There are three principal types of hose used for fire fighting operations — woven-jacket, rubber lined; wrapped or braided in plies; and unlined.

Woven-jacket, Rubber Lined

This type of hose consists of a lining or tubing made of natural or synthetic rubber covered by a woven jacket of cotton and synthetic fibers. It is available with a single jacket or with multiple woven jackets. The single jacket type is generally used in connection with industrial yard hydrants and standpipe systems, while multiple jacket hose is used by the public fire service. The single jacket type is found most commonly in 1½- and 2½-inch (37- and 62-millimeter) sizes. Multiple jacket hose comes in a number of sizes from 1½ to 6 inches (37 to 150 millimeters). Hand lines are usually 1½- and 2½-inch (37- and 62-millimeter) hoses. Three-inch (75-millimeter) hose is used to supply heavy stream appliances, while 4- to 6-inch (100- to 150-millimeter) hose supplies fire pumps from hydrants.

Wrapped or Braided in Plies

Wrapped or braided hose consists of a rubber tube, reinforced by a wrapped or braided fabric, and an outer rubber cover. Such hose is called "booster hose" and is used in ¾- and 1-inch (19- and 25-millimeter) sizes by public fire departments. Smaller diameter hoses of similar construction are used on portable fire extinguishers. Larger hose of this type, which contains a spiral wire, is used by fire departments to supply their pumps by drafting water from a lake, stream, or reservoir.

Unlined Hose

Unlined hose consists of a fabric tube, usually made of linen. When the fabric becomes wet, the threads swell and make the tube watertight. This type of hose is used at indoor standpipe locations where it can be kept dry when not in use. It is available in sizes ranging from 1¼ to 2½ inches (31 to 62 millimeters).

HOSE COUPLINGS

There are several types of couplings used on fire hose. The most commonly used is a threaded coupling with a stationary male end and a female end that swivels

Figure 4-1. Three-piece 2½-inch hose coupling showing details of standard design. The hose is held in the ends by metal ferrules expanded inside the ends. Male threads can be seen. Female thread is in the central swivel piece.

(Elkhart Brass Mfg. Co., Elkhart, Ind.)

(Akron Brass Co., Wooster, Ohio)

Figure 4-2. A comparison of pin-lug (left) and rocker-lug (right) hose couplings.

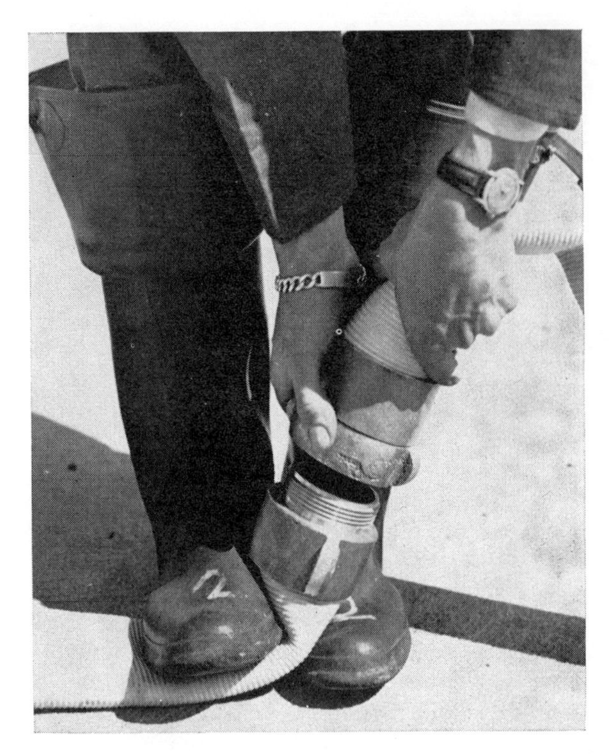

Figure 4-3. Making and breaking couplings by one fire fighter with hose butt held between feet.

Figure 4-4. Making and breaking coupling by one fire fighter with hose supported by thigh.

on a third piece, which is fastened to the hose. A gasket fits snugly inside the female end to prevent leakage. Threaded couplings are designed to be made up hand tight.

Care should be taken to ensure that couplings are not dropped or ridden over by vehicles. They could become distorted or "out-of-round." It is good practice to examine each coupling after each use. Hose lengths with damaged couplings should be set aside for repair or replacement.

NOZZLES

Nozzles are for shaping, directing, and controlling the flow of fire hose streams. There are several types of nozzles, including those which produce straight streams and water sprays. In addition, there are a number of special purpose types.

Straight stream nozzles for use on 2½-inch (62-millimeter) hose generally consist of a tapered playpipe with 2½-inch (62-millimeter) threads at the inlet and 1½-inch (37-millimeter) threads at the outlet, a ball valve to turn the stream on and off, and a nozzle tip. Interchangeable tips provide different flow rates at a given pump pressure.

There are three general types of spray nozzles for fire hose streams. One is a fixed pattern spray nozzle, usually equipped with a shutoff valve. This can be used wherever it is desirable to avoid the accidental application of a straight stream to the fire.

Another type of spray nozzle gives the operator continuous control in adjusting between a straight stream and a wide angle spray. In some, the discharge varies as the spray pattern changes. Newer models give nearly constant discharge over a wide range of spray patterns. Some of these adjustable nozzles have a "closed" position on the pattern selector, while others have a positive shutoff valve.

A third type of spray nozzle provides either a straight stream or a fixed pattern spray. A three-position valve directs the flow through the straight stream orifice or into the spray pattern chamber or shuts off the flow. An extension or applicator may be attached to some nozzles of this type for reaching out over flammable liquid fires.

One type of special nozzle is the cellar or distributor nozzle for 2½-inch (62-millimeter) hose. Water enters a swivel-mounted chamber that has several solid stream orifices pointing in different directions. Forces created when the water enters the chamber cause the chamber to spin. The result is a coarse spray. Such nozzles are useful against cellar fires when fire fighters cannot gain access to the cellar. If a hole is chopped in the floor above, the hose line and nozzle can be fed into the cellar. The nozzle is not equipped with a shutoff valve, but one should be installed somewhere in the line.

(Elkhart Brass Mfg. Co., Elkhart, Ind.)

Figure 4-5. A solid stream nozzle with shutoff and 1½-inch threads connected to a 2½-inch playpipe with ladder hook.

(Elkhart Brass Mfg. Co., Elkhart, Ind.)

Figure 4-7. An adjustable fog nozzle with shutoff and 1½-inch threads connected to a 2½-inch playpipe with ladder hook.

HOSE APPLIANCES

There are several accessory items available for use with industrial fire hoses. They include valves, adapters, reducers, double-ended connections, siamese and wye connections, hose clamps, and hose jackets.

Yard hydrants in industrial properties are equipped with 2½-inch (62-millimeter) outlets for direct connection of hose lines. Such hydrants are not intended to supply fire department pumpers. Often, these hydrants are equipped with individual valves so that each line can be controlled independently. The valves may be an integral part of the hydrant assembly, bolted to the hydrant itself, or they may be threaded on the outlets themselves. They may be screw type gate valves or quarter turn ball valves. In either case, they will have a female swivel coupling on the inlet side and a male coupling on the outlet side.

Adapters are used to connect different types of threads of the same diameter. One example of their use is in connecting a hose having a fire department thread to a standpipe having a pipe thread.

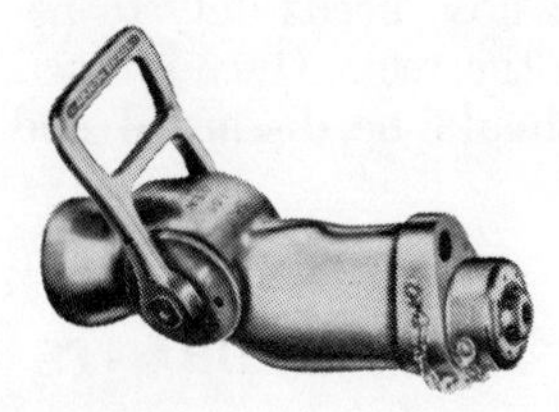

(Akron Brass, Wooster, Ohio)

Figure 4-6. A 1½-inch 4-NAP fog nozzle, a combination fixed pattern fog and solid stream nozzle, also available in 1- and 2½-inch sizes.

Reducers or reducing couplings are used to extend a hose line with a smaller diameter hose. They are also used to connect a hose to a larger diameter pump or hydrant outlet.

Double-ended connections may be either double male or double female. Double males are used to connect two female swivel couplings. A double female is a device having two female swivel couplings to permit connection of two male couplings.

A siamese is a device for using two supply lines to feed a single hose stream. It has a female swivel coupling at each inlet and a male coupling at the outlet. It may be a plain waterway or it may contain a clapper valve (check valve) in each inlet branch to prevent the escape of water from an unused inlet. Water pressure inside the siamese automatically closes the valve on the unused inlet.

The purpose of a wye connection is to permit a single supply line to feed two hose streams. It has a female swivel coupling at the inlet and a male coupling at each outlet. It is advisable to attach a valve —

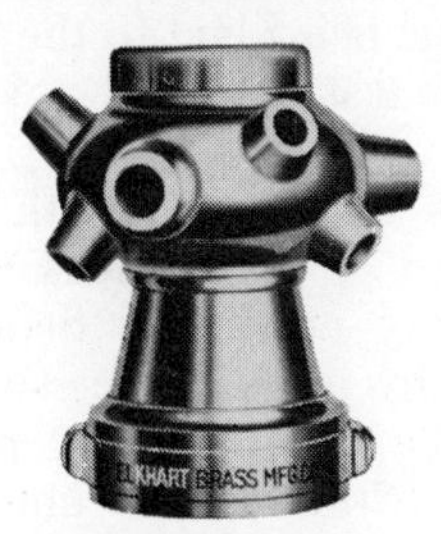

(Elkhart Brass Mfg. Co., Elkhart, Ind.)

Figure 4-8. A distributor-type cellar nozzle.

(M & H Valve & Fittings Co., Birmingham, Ala.)

Figure 4-9. Hydrants for use on private yard systems of industrial plants usually have only 2½-inch outlets, as they are not intended to supply fire department pumpers. It is good practice to have a gate on each outlet.

either a gate valve or a quarter turn ball valve — to each outlet. A reducing wye permits the use of a large line to feed two smaller hose streams.

A hose jacket is a device intended to be clamped around a leak in a 2½-inch (62-millimeter) hose, so that the hose can be continued to be used temporarily until it is safe or convenient to replace the defective length.

STORING HOSE

For efficient fire fighting, it is essential that hose be stored for use in an accessible place and in a manner that will permit its being laid to the fire in the shortest time and with the least effort.

Where hose is to be used with indoor standpipe systems, approved or listed racks and reels for hose storage are usually provided. Semiautomatic racks, intended for use by untrained personnel, are designed to store 50 to 100 feet (15 to 30 meters) of 1¼- to 2½-inch (31- to 62-millimeter) unlined linen hose or lightweight, woven-jacketed, lined hose. Nonautomatic racks for the storage of 100 to 200 feet (30 to

(Elkhart Brass Mfg. Co., Elkhart, Ind.)

Figure 4-10. Valves with female swivel coupling for use on hydrants and in hose lines — screw-type gate valve (left) and quarter-turn gate valve (right).

60 meters) of unlined 2½-inch (62-millimeter) hose are intended for use by trained fire fighters.

Reels are also available for the storage of ¾- to 1½-inch (19- to 37-millimeter) rubber hose. In this case, it is not necessary to remove all the hose from the reel to obtain full flow.

For outdoor hose storage, hose houses should be provided in most plants. A house may be built over and around or near each hydrant. It should contain 100 to 200 feet (30 to 60 meters) of 2½-inch (62-millimeter) hose connected to a hydrant and equipped with a nozzle. Additional hose may be stored in hose houses in 50-foot (15-meter) rolls. The hose house should be equipped with wrenches, axes, and other tools as may be needed.

In some plants, a supply of 2½-inch (62-millimeter) hose may be carried in a hand cart or on a hose reel. An industrial plant covering a wide area may have motorized fire trucks to carry hose and other fire fighting equipment. When hose is rolled for storage in a hose house, it is recommended that the donut roll be used. The convenience of the donut roll becomes evident when it is carried and when it is unwound.

Woven-jacketed, lined hose should be swept clean or washed, whichever is necessary, after each use and thoroughly dried before being returned to storage. Rubber covered hose needs only to be wiped clean as it is rerolled on the reel. Once unlined linen hose has been used, it should be discarded and replaced with new hose.

USING HOSE FROM HYDRANTS

Water leaving a fire hose nozzle develops a counter-thrust known as nozzle reaction. If the hose crew is not prepared to deal with this reaction, it may lose

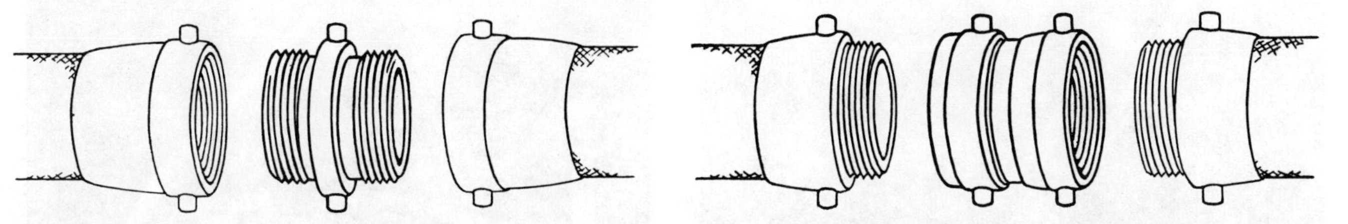

Figure 4-11. Double male hose connection.

Figure 4-12. Double female or swivel connections.

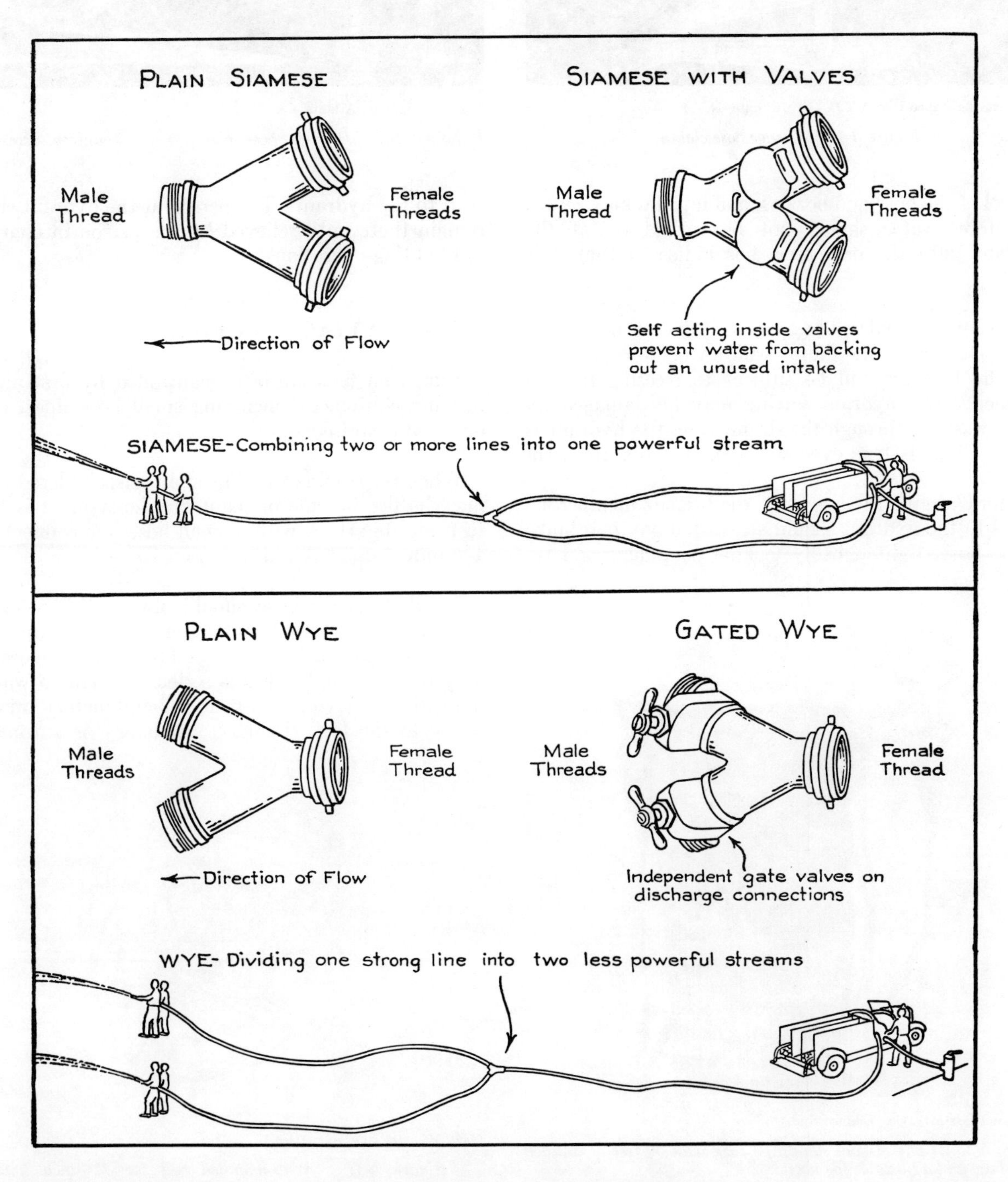

Figure 4-13. Siamese and wye connections.

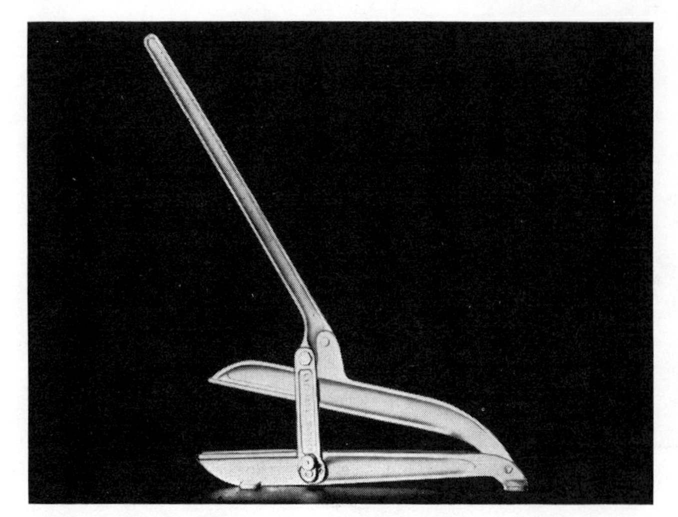

(American LaFrance Div. A.T.O. Corp, Elmira, N.Y.)

Figure 4-14. Warner hose clamp.

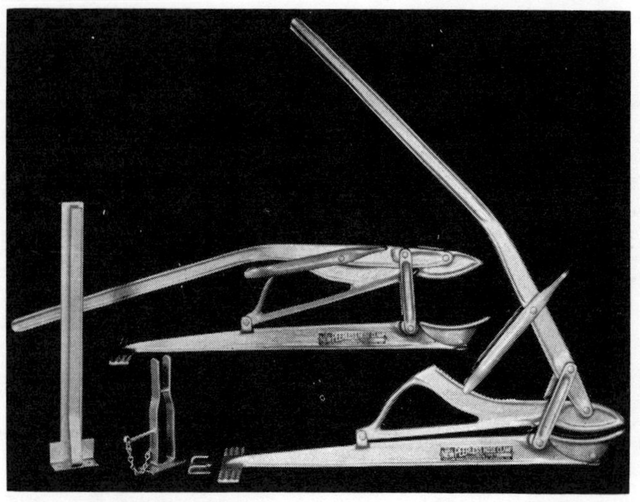

(Kellan Mfg. Co., Atlanta, Ga.)

Figure 4-16. Peerless hose clamp with mounting accessories.

control of the hose and nozzles, and injuries may result. Therefore, water should not be turned on at the hydrant until the hose crew has indicated that it is ready.

Whenever a hydrant is opened, it should be opened fully. Otherwise, the drains will not operate properly, and the hydrant will be subject to freezing in cold weather. The hydrant setting may be damaged by water washing through the drain while the hydrant is in use. This is true even when small flows are used.

After charging the hose line, the brigade member assigned to the hydrant should straighten any bad kinks in the hose, tighten badly leaking couplings, and return to the hydrant. The person at the hydrant should remain there until relieved by the person in charge of fire fighting operations.

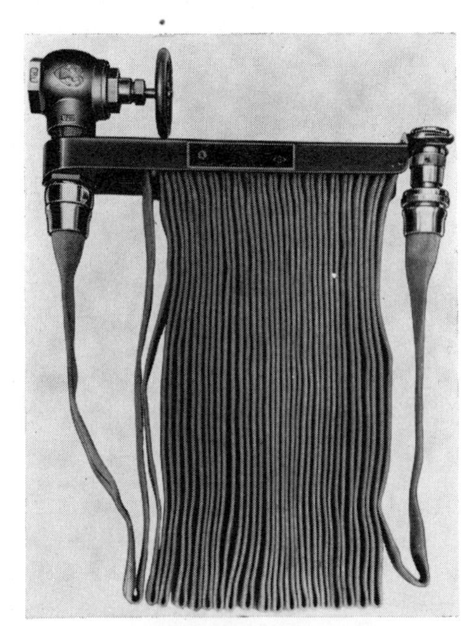

(Elkhart Brass Mfg. Co., Elkhart, Ind.)

Figure 4-15. Conventional standpipe rack with 50 feet of unlined linen hose and adjustable fog nozzle.

INITIAL ATTACK ON FIRE

Industrial fires are often controlled by first aid fire fighting equipment, including small hose lines, or by automatic sprinklers.

When responding to a fire in an unsprinklered building, the fire brigade or plant fire department is likely to begin its attack with a small hose line connected to a standpipe and stored on a rack or reel.

Small diameter hose should always be backed up by a 2½-inch (62-millimeter), or larger, hydrant hose line. The person in charge of fire fighting operations may decide that the fire is serious enough to warrant immediately laying 2½-inch (62-millimeter) lines directly to the fire. Usually, stationary or automotive

(Seco Mfg. Inc., Detroit, Mich.)

Figure 4-17. Wall-mounted reel for 1½-inch hose.

(Powhatan Brass & Iron Works
Div. of A.T.O. Corp., Ranson, W. Va.)

Figure 4-18. Water may be fed to the rubber hose on this reel without having to remove all the hose from the reel.

fire pumps are provided to ensure adequate pressure in the water distribution system to supply the lines.

Most fires in sprinklered buildings can be controlled by streams from 1½-inch (37-millimeter) hose. Larger hose lines should not be connected to hydrants on mains that are feeding sprinkler systems unless a fire department pumper is used to reinforce the sprinkler system's water supply. Some sprinkler systems have provisions for connecting 2½-inch (62-millimeter) hose lines to sprinkler risers. Where such connections exist, they are for fire department use only, and should be used only when a fire department pumper is supplying additional water to the sprinkler system.

ADVANCING HOSE LINES

Advancing a hose line, even a 2½-inch (62-millimeter) line, is easy to do when it is taken from a hose house, because hose houses are generally located to permit short hydrant-to-fire layouts. Customarily, a predetermined number of hose lengths are connected together with a nozzle attached to the last length. Extra hose may be rolled and also stored in the hose house.

Fire in remote areas of a sprawling or multistory building may necessitate the use of additional lengths

(Star Supply Corp., Columbus, Ohio)

Figure 4-19. Hose house of compact dimensions for installation over a yard hydrant may be of steel or aluminum construction.

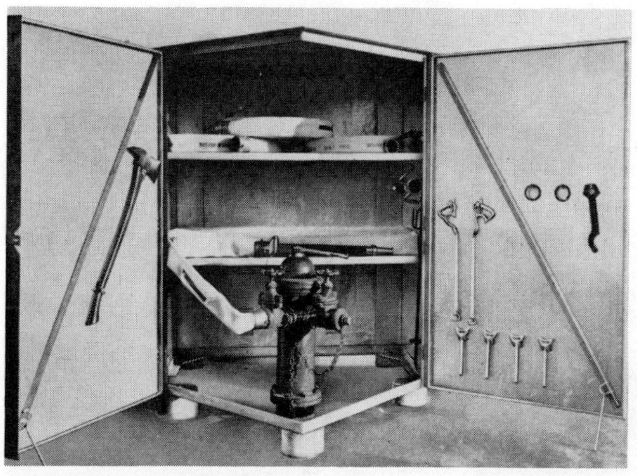

(Grinnell Co., Providence, R.I.
W. D. Allen Mfg. Co., Broadview, Ill.)

Figure 4-20. Hose house of five-sided design for installation over a yard hydrant may be made of wood or steel equipped with a floor.

of hose. Serious delays will result if the hose line is laid out and charged, and then has to be shut down so that more hose can be added. Therefore, fire brigade members should learn how much working line is needed to reach various out-of-the-way places in the plant. This can best be accomplished by including hose layouts in training sessions.

CARRYING HOSE

Dragging hose by brute strength is difficult and often slow work. There are, however, various methods of carrying hose that make the work of laying lines quicker and easier.

For very short distances, several folds of hose may be dragged under the arm. Another method is to carry hose on the shoulder.

This carry has many uses on the fire ground. It is

Figure 4-21. The first step in making a donut roll by the two-man method. Male coupling is 4 to 5 feet behind the swivel coupling.

Figure 4-22. Second step in making a donut roll by the two-man method. Hose is rolled toward the couplings.

Figure 4-25. Picking up a completed donut roll.

Figure 4-23. Start of a donut roll by the one-man method.

Figure 4-26. Loading shoulder carry for 2½-inch hose.

Figure 4-24. Completion of a donut roll by the one-man method.

a convenient method of advancing lines up a stairway so that they will not snag. With the shoulder load, each brigade member advances until the hose tightens behind him, then he lets it pay out as he walks. The last one to let his hose pay out is the one carrying the nozzle.

Where the line must later be extended, donut rolls can be attached or 1- or 1½-inch (25- or 37-millimeter) hose can be attached to the 2½-inch (62-millimeter) nozzle, using the proper fittings.

Where extra hose is kept in rolls, as in a hose house, each roll should be strapped or kept in a sling that can be loaded on the back. This saves effort and prevents the roll from becoming loose or tangled.

Figure 4-27. Method of advancing a 2½-inch hose line in a straight lay. Pipeman carries hose across chest and nozzle over the shoulder. Other men put hose over shoulders, couplings at chest. This automatically spaces men so that there is enough slack hose on the ground for convenient maneuvering.

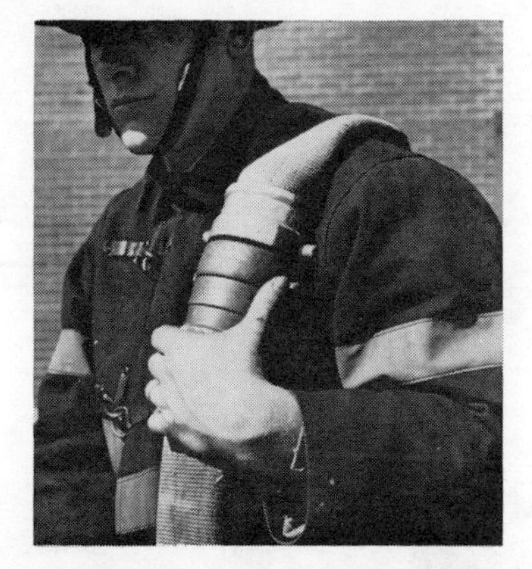

Figure 4-28. How hose is carried at hose couplings in making a straight hose lay.

Figure 4-29. How a nozzle is carried by lead man in making a straight hose lay.

Figure 4-30. Lines should not be charged until hose crews are ready. Here they are holding the nozzle on a 2½-inch line. The hose is lashed back of the playpipe with a rope hose tool, a loop of which can be seen over shoulder of fire fighter on left.

Figure 4-31. Advancing a dry line up a stairway. When line from outside is taut, it starts peeling off the shoulder of the last man as he advances. When his load is laid, he calls out, and then the next man allows the line to feed off his shoulder. Lead man is last to lay his load out.

Figure 4-32. Sequence in advancing a dry 2½-inch line of hose up a ladder. Slack hose at base of ladder should be kept clear of tormenter poles.

Figure 4-33. Sequence in advancing a dry 2½-inch line of hose up a ladder.

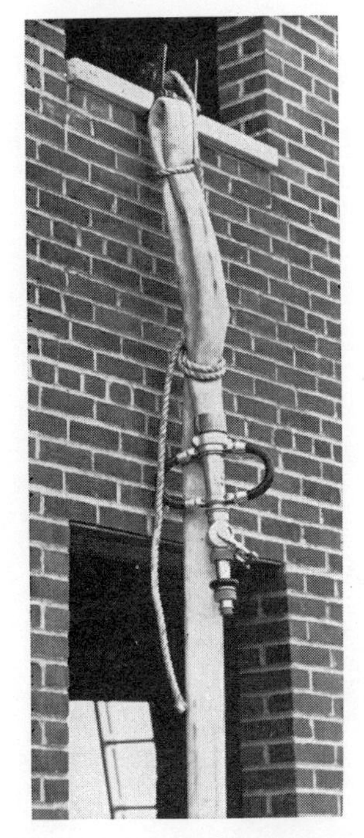

Figure 4-34. Hoisting dry line of hose with nozzle attached. Hitches are in center of line on opposite side from nozzle. Nozzle points down to protect tip.

Figure 4-35. Lowering dry line of hose with nozzle attached. Nozzle pointing opposite to direction of travel to protect tip. Shutoff or controlling valve secured by rope.

HOISTING HOSE WITH A HAND LINE

At times, it is necessary to take a hose line through an upper story window or over a roof parapet. This can be done with a rope line. Drop the line from the desired point and secure it in the following manner. Fold the hose back on itself and tie it so that the nozzle faces downward and away from the building as the hose is hoisted. This prevents the nozzle from being bumped against the building and possibly damaged. Lower the hose line with the nozzle pointing upward and the nozzle valve secured in the closed position by the rope hand line. For ease in hoisting, the line should not be charged. Use a hose roller to protect the hose from abrasion where it is pulled over a sill, eaves, or coping.

REPLACING BURST SECTIONS OR EXTENDING LINES

If it becomes necessary to replace lengths of hose in, or add lengths of hose to, a charged line, it should be done quickly. Bring up the required amount of hose, and lay it in place. Shut down the line, make the necessary connections, and recharge the line. If it is not convenient to shut down the line at the hydrant or pump, use a hose clamp or kink the line to stop the flow of water.

Suggested Reading

Fire Service Training Committee, *Handling Hose and Ladders*, 1969, NFPA, Boston.

NFPA 14, *Standard for the Installation of Standpipe and Hose Systems*, 1976, NFPA, Boston.

NFPA 24, *Standard for Outside Protection*, 1973, NFPA, Boston.

NFPA 1901, *Standard for Automotive Fire Apparatus*, 1975, NFPA, Boston.

Fire Hose Practices, 6th Edition, 1974, International Fire Service Training Association, Stillwater, Oklahoma.

Chapter 5

HANDLING LADDERS

Where ladders are provided as part of the fire brigade's emergency equipment, brigade members should be skilled in their use. Proper procedures for carrying, raising, and climbing ladders should be practiced until they become almost automatic.

DESIGN, TYPES, AND PRECAUTIONS

In order to understand the safe and efficient use of ladders, one must know something about their construction, advantages, and limitations.

Fire service ladders are made of either wood or metal. Most long ladders are made of metal. The principal parts of a ladder are its beams — or sides — and its rungs. Ladder beams are either solid or trussed. Trussed construction is used in long ladders to make them strong enough to withstand the loads imposed by fire fighting service. Rungs are round bars of sufficient diameter to provide the required strength and to be easily grasped. Metal rungs are tubular and may be fluted horizontally to provide for a sure grip and firm footing.

Figure 5-1. Solid beam ladder.

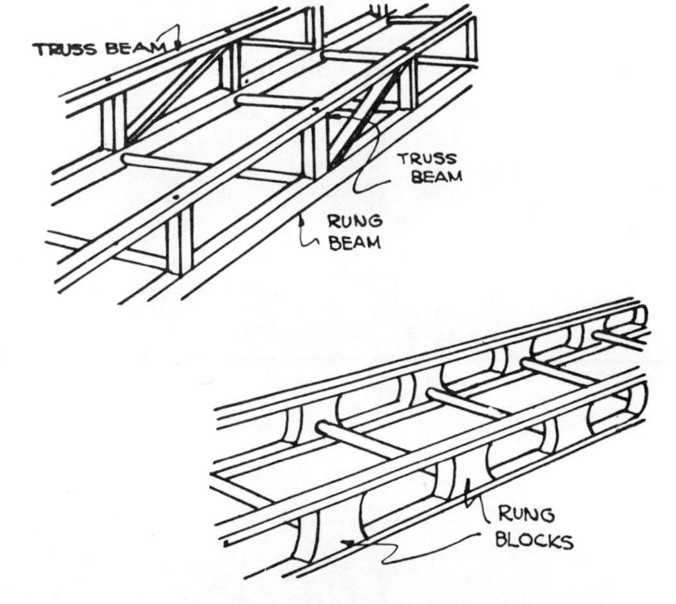

Figure 5-2. Trussed beam ladder.

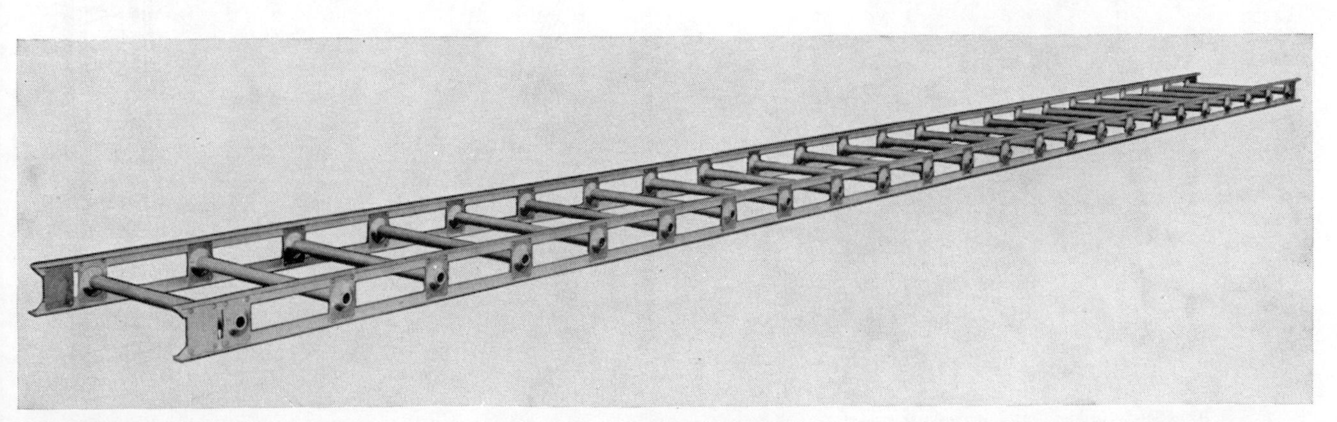

(Aluminum Ladder Co., Florence, S.C.)

Figure 5-3. Straight or wall ladder.

Lightweight aluminum alloy ladders have come into general use. When constructed as a three-section extension ladder, a 45- or 50-foot (13.7- or 15-meter) ladder can be carried on a fire brigade pumper truck.

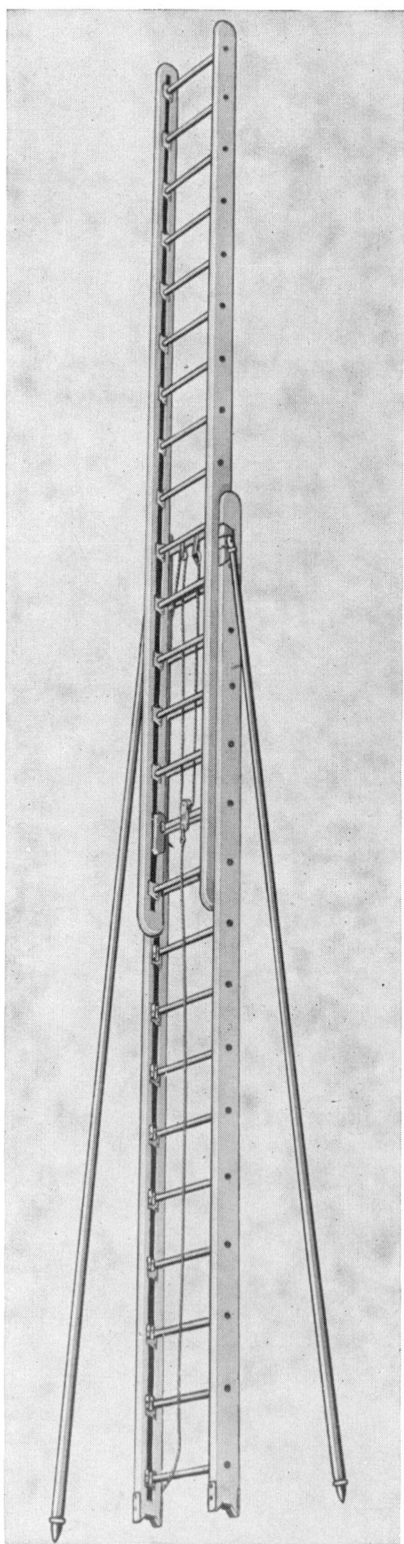

(Duo-Safety Ladder Corp., Oshkosh, Wis.)

Figure 5-4. Two-section fire ladder with tormenter poles.

The fire brigade member should be familiar with some of the terms that are applied to commonly used ladders. They are the following:

- Bed ladder — the lowest section of an extension ladder;
- Fly ladder — the upper sections of an extension ladder;
- Butt — the ground end of a ladder;
- Heel — the extreme ground end of a ladder beam;
- Halyard or fly rope — the rope for raising the fly; and
- Pawl or dog — the mechanism on the lower end of the fly ladder that secures it to the bed ladder and holds the rungs at a desired elevation.

Straight ladders, sometimes called wall ladders, are made in one section only. They are made with either solid beam or trussed beam construction in standard lengths running from 10 to 28 feet (3 to 8.5 meters). A few straight ladders 30 feet (9 meters) or more in length are used.

Extension ladders, as their name implies, are generally made of two or three sections. Fly ladders slide through guides on the upper end of the bed ladder. The lower end of each fly ladder is equipped with locks that hook over the rungs of the bed ladder to keep the extended fly ladder in position. On all except the very short ladders, the fly is raised with a halyard that is fastened to the lower rung and operates over a pulley attached to the upper end of the bed ladder. Extension ladders are made in lengths from 14 feet (4.2 meters) to 50 feet (15 meters) and longer.

Figure 5-5. Baby extension ladder, 14 feet.

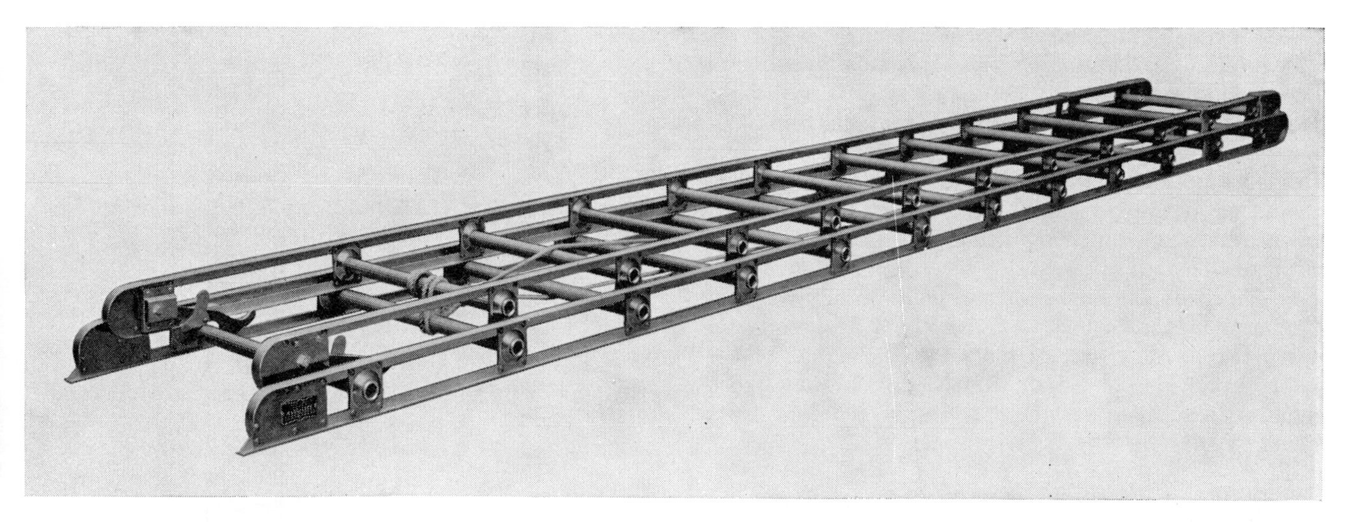

(Aluminum Ladder Co., Florence, S.C.)

Figure 5-6. Two-section extension ladder.

(Aluminum Ladder Co., Florence, S.C.)

Figure 5-7. Two-section extension ladder.

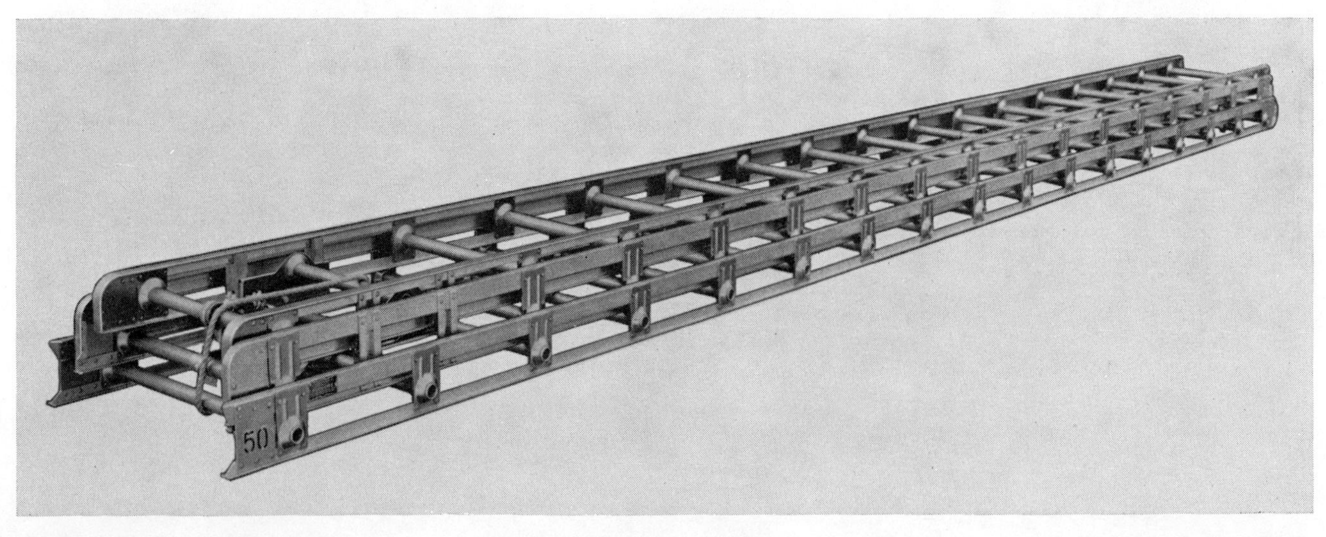

(Aluminum Ladder Co., Florence, S.C.)

Figure 5-8. Three-section extension ladder.

The primary purpose of a roof ladder is to provide a firm foothold on a sloping roof. It is an adaptation of a wall ladder. Basically, the roof ladder is a straight ladder with a pair of hooks mounted at the top. The hooks are placed over roof peaks, sills, walls, or coping to secure the top of the ladder. On many roof ladders, mounting hardware permits the hooks to be turned inward between the beams and set so that they do not protrude beyond the ends of the beams when not in use. To ready the hooks for use, pull them out beyond the ends of the beams and give them a one-quarter turn so that they lock in place. Roof ladders of the hook type range in length from 10 to 20 feet (3 to 6 meters), and may be either the solid beam or truss beam type.

The 10-foot (3-meter) collapsible ladder is especially useful for inside work. It can be easily carried up stairways and to places where the ordinary ladder could be cumbersome to maneuver. To operate the ladder, place it in a vertical position, grasp both beams, and push them apart.

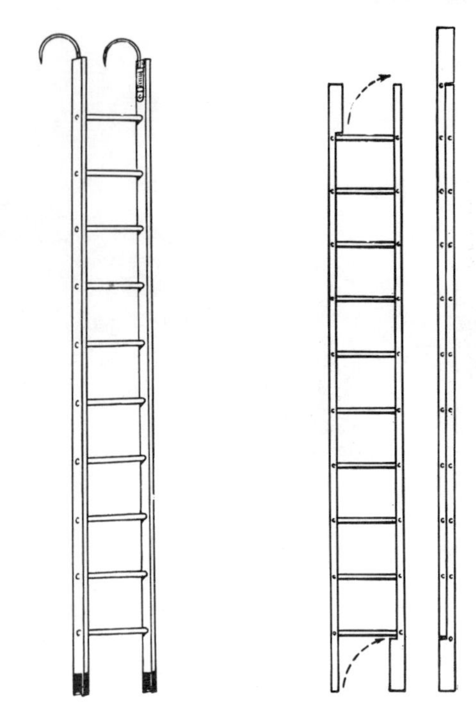

Figure 5-9. Wooden roof ladder (left) and collapsible attic ladder (right).

The practices taught in this manual are limited to the ladders commonly found on a fire department pumper or hung in convenient places in an industrial plant. They include the straight ladder, the extension ladder, and the roof ladder. Manufacturers' instructions should be followed for special techniques required with large or special ladder equipment.

Ladders must be maintained in first-class condition. The manufacturers have selected each part with care. The wood has been properly aged and seasoned, or the metal has been properly alloyed to stand up under rigorous use. In time, however, wood fibers may become hard and lose their elasticity. Fibers may become damaged. Beams may splinter, and rungs may become loose if periodic maintenance is not performed. Weakening of a ladder from any of these causes or from overloading may cause ladder failure. The loosening of rungs or blocks presents additional hazard and usually results from neglect.

Raising a ladder on its beams tends to strain it unnecessarily and to loosen its rungs. Rolling a ladder

Figure 5-10. One-man carry for short ladder.

beam over beam, "walking" a ladder, or sliding a ladder along a cornice to a new position are considered bad practices. If they are ever so treated, used for forcible entry, or subjected to any unusual load, ladders should be inspected and tested afterward.

The greatest stress on fire ladders usually results from overloading them. Ladders are commonly overloaded when they are placed at an improper angle. The butt of the ladder usually should be out from the wall 20 to 25 percent of the height. About 1 foot (0.3 meter) out for 4 or 5 feet (1.2 or 1.5 meters) of height is a safe distance for trussed ladder. Solid beam ladders should be even straighter. Ladders are also overloaded when too many people crowd to the upper end, especially when there is additional stress from charged hose lines operating from the ladder. The number of persons on a ladder at one time should be the minimum number actually needed to do the job at hand. One brigade member should be stationed on the ground at the base of the ladder to secure the ladder while it is in use. At a smoky fire, brigade members should take turns operating at the top of the ladder, with not more than two of them near the top at one time.

Figure 5-11. One-man ladder raise.

USING STRAIGHT LADDERS

Ladder practices include the job of taking the ladder from its resting place, carrying it to the location where it is needed, placing it, raising it to the climbing position, securing it, climbing it or using it, lowering it, and carrying it back to the place where it is kept and depositing it there. For this study, the practices are considered in the following sequence:

(*a*) Securing the ladder,
(*b*) Carrying the ladder,
(*c*) Placing the ladder,
(*d*) Raising the ladder,
(*e*) Lowering the ladder, and
(*f*) Climbing the ladder.

Lightweight ladders 18 feet (5.5 meters) long or less can be handled safely by one person. Those 20 feet (6 meters) or longer should be handled by two people.

The procedures for one person to handle a short ladder are as follows:

(*a*) Grasp the ladder near its center of balance, lift it, and extend one arm between the rungs so that the center of balance rests on your shoulder. Secure the ladder by grasping a rung with the hand that you passed between the beams. Carry the ladder heel first and tilted slightly downward in front of you so that it does not interfere with your line of vision or snag on overhead obstructions.

(*b*) When at the desired location, place the heel of the ladder against the building or structure.

(*c*) Raise the ladder by walking toward the foot of it while grasping progressively lower rungs.

(*d*) Grasp the ladder with the hands about three rungs apart and set it at a climbing angle. A good general rule to follow is to place the foot of the ladder away from the building at a distance that is about one-fourth of the height to which the ladder is extended.

(*e*) To lower the ladder, reverse the preceding steps.

(*f*) When lowered to shoulder height, the ladder can be placed on the shoulder and carried to the desired storage place.

The procedures for two people to handle a longer ladder are as follows:

(*a*) If the ladder is on the ground, take a position near the end — one brigade member at the foot and the other at the top. Face the direction opposite to that in which you intend to travel. Palm up, grasp the second rung from the end and lift the ladder to shoulder height, turning as the ladder is raised and thrusting your free arm through the ladder. With the hand that you passed between the beams, grasp the forward rung or the upper beam. If the ladder is mounted on a truck, it can be loaded directly on the brigade members' shoulders.

(*b*) To transport the ladder, walk in step and use your free hands to guard against traffic and obstacles.

(*c*) Reverse the foregoing procedure to lower the ladder to the ground.

(*d*) There are two basic ladder raises — the flat raise and the beam raise. The flat raise begins with both beams of the ladder resting on the ground. If your position is at the foot of the ladder, place one foot on the heel of each beam, crouch, reach forward, and grasp a convenient rung. If you are going to raise the ladder, take a position alongside the ladder, facing the top, and grasp a rung. Lift the ladder, pivot and turn under it, and step forward toward your partner at the heel of the ladder while grasping progressively lower rungs.

(*e*) If the ladder has been raised from a position perpendicular to the wall, it may be lowered in place against the building. If it was raised from a position parallel to the wall, it must be pivoted on one beam and lowered in place against the wall.

(*f*) The beam raise begins with one beam of the ladder resting on the ground. If you are at the bottom end

Figure 5-12. Two-man ladder carry.

Figure 5-13. Two-man ladder carry.

of the ladder, place one foot on the heel of the lower beam, and grasp the upper beam with two hands. If you are at the top end of the ladder, grasp the center of the fourth or fifth rung, lift the ladder, and pivot under it. Walk toward your partner, pushing the ladder toward the vertical as you go. Your partner will steady the ladder with one hand and help you raise it with the other.

(*g*) To lower the ladder, reverse the procedure.

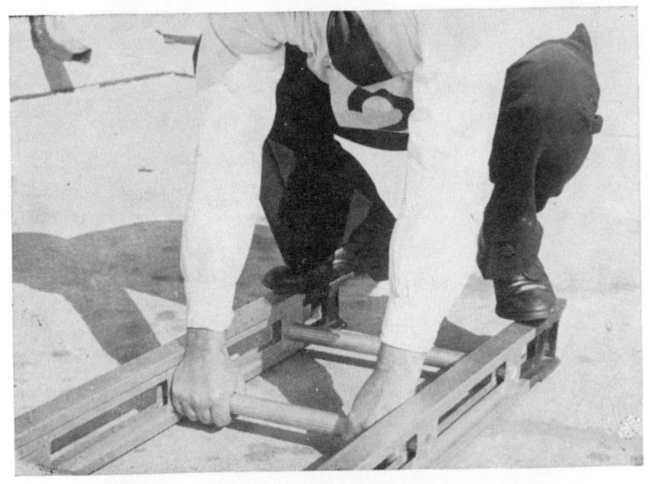

Figure 5-14. Two-man flat raise, position of heel man.

Figure 5-15. Two-man flat raise.

Figure 5-16. Two-man beam raise, first step.

USING EXTENSION LADDERS

If an extension ladder is no longer than 24 feet (7.3 meters), one person can carry and raise it. Place the ladder against the building in the same manner as was described for a straight ladder. Place one foot on the bottom rung and pull the top of the ladder away from the building. Do not let the ladder go beyond the vertical and lean over your head, because you can easily lose control of it.

With the ladder tilted slightly toward the building, grasp the halyard and pull to extend the fly. After the fly has been raised, secure the halyard and lower the ladder against the building. Adjust the space between the building and the bottom of the ladder to be approximately one-quarter of the vertical distance from the bottom of the ladder to the top.

Figure 5-17. Two-man beam raise, second step.

To lower the ladder, place one foot on the bottom rung, bring the ladder to the vertical position, and reverse the preceding steps.

If the ladder is mounted on a truck or hanger, you can load it directly on your shoulder. Release the ladder locks, hold the ladder, and determine its center of balance. Lift it off the hangers and proceed as described.

If the ladder is to be set on concrete, it may be necessary to butt the heel against the base of the building or some other firmly anchored object before raising it. Then set the ladder at the desired distance from the building before extending the fly.

If the ladder is to be placed in a window, it should be placed at one side, not in the center.

Extension ladders from 24 to 28 feet (7.3 to 8.5 meters) long can be raised by two people. With the ladder on the ground, the procedure is the same as for a straight ladder.

Place the ladder on the ground at the desired location with the beams parallel to the building. After the ladder has been raised, pivot it on one beam so that the rungs are parallel to the wall. If you are on the fly side of the ladder, it is your job to steady the ladder while your partner, who is on the other side, pulls on the halyard to extend the fly. Once the fly has been extended to the desired height, together lower the ladder against the building.

To lower the ladder, reverse the procedure.

If you are supporting the ladder while the fly is being raised, be sure that your hands and feet are not in a position where they can be caught and injured by the fly ladder.

If you are heeling the ladder while it is being raised, you can aid materially by assuming a crouched position and throwing your weight backward to balance the ladder. Keep your eye on the top of the ladder while it is being raised.

If the ladder is mounted on a truck or on hooks, it may be loaded directly onto the brigade members' shoulders. It can be unloaded in the same way.

Longer extension ladders — 30 to 40 feet (9 to 12 meters) — require from three to six people to handle. When three people are used, one heels the ladder while the other two lift the ladder, pivot under the beams, and push the ladder up to vertical.

Figure 5-18. Raising and placing short extension ladder.

Figure 5-20. Two-man raise, short extension ladder.

Figure 5-19. Raising fly on short extension ladder.

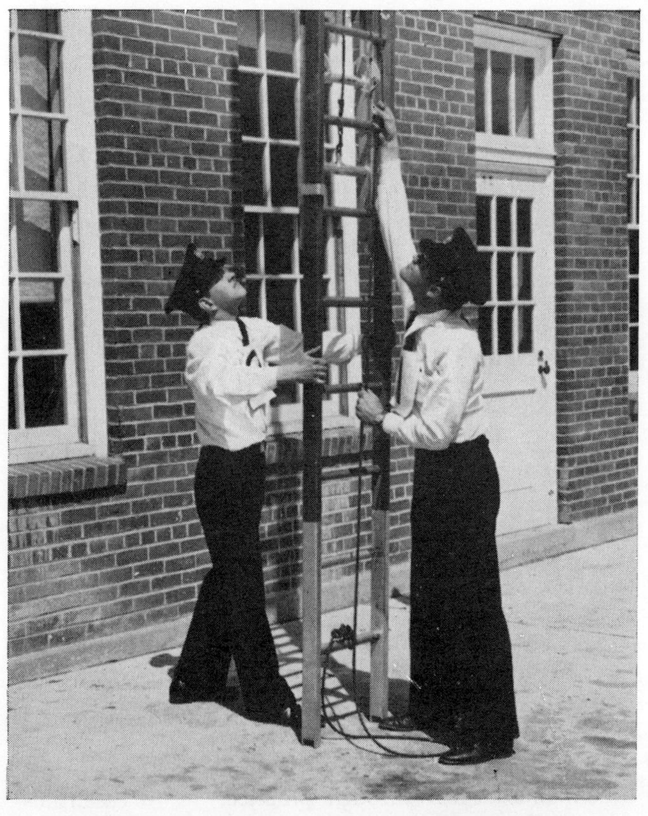

Figure 5-21. Two-man raise, short extension ladder.

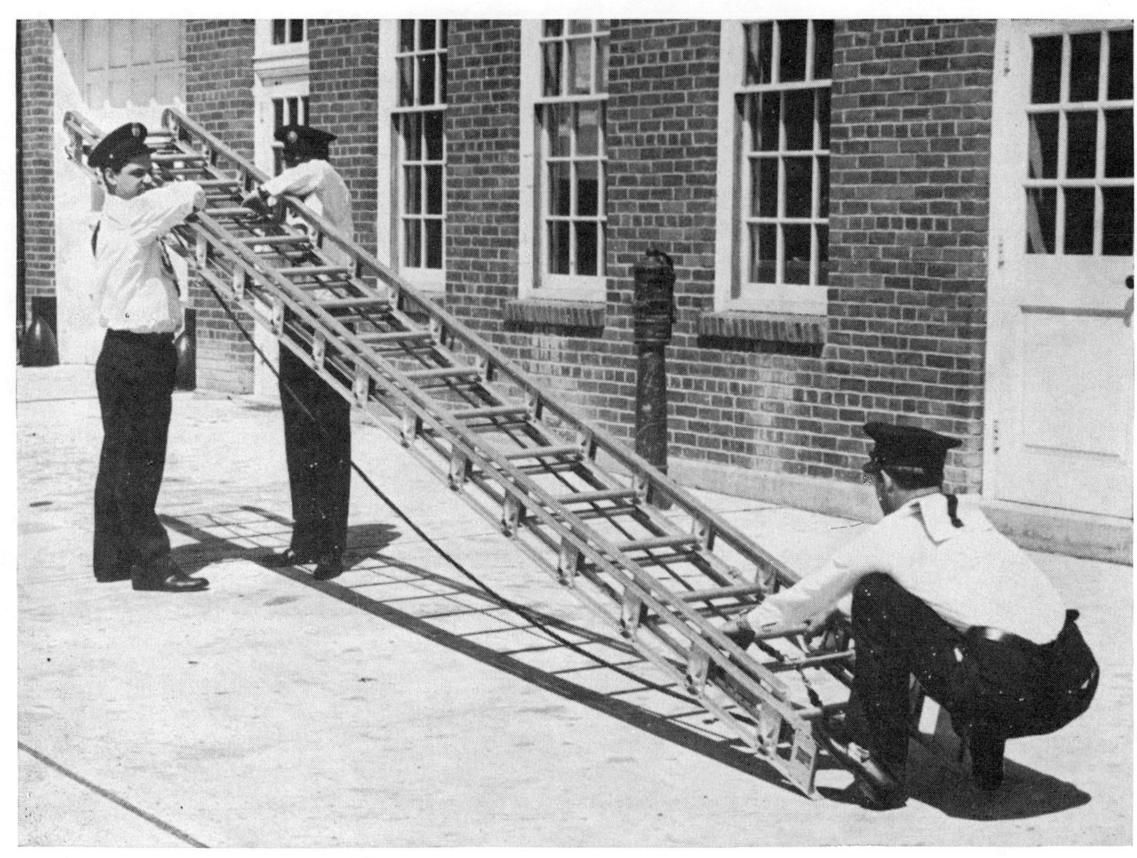

Figure 5-22. Extension ladder, three-man raise.

USING ROOF LADDERS

A roof ladder is usually carried up another ladder to be used. Practice evolutions consist of carrying the roof ladder to, and up, the other ladder and placing it in position on the roof. Carry a short roof ladder the same way you would carry a short straight ladder — the top end forward and tilted downward slightly.

Open the hooks before climbing the main ladder and carry the roof ladder up the main ladder with the hooks turned outward. With the hooks pointed downward, push the roof ladder up the roof until the hooks are secured over the ridge pole or peak.

A longer roof ladder will require two people to get it into position. Both will carry the roof ladder up the main ladder until the first reaches the eave and locks a leg around the main ladder rung and beam. The first person on the main ladder pushes the roof ladder up until the hooks are secured over the peak, while the second brigade member assists from a position lower on the main ladder.

CLIMBING LADDERS

Climbing a ladder comes naturally to some, while others must learn through practice. Two essentials in climbing are rhythm and safety. For most people, the best rhythm is obtained by taking every rung with the feet and hands. Some prefer taking every other rung with the hands, but this may result in the ladder beginning to vibrate or the climber missing a rung in the dark, which could prove hazardous. However, climbing habits are something that will vary with the stature of the individual.

Keep both hands on the rungs except when carrying something, in which case slide your free hand up the beam so that you have a hold of the ladder at all times. Practice taking a firm, quick hold on each rung. This will safeguard you in the event your feet slip. To prevent wobbling, place your feet at the center of the ladder. Keep your body nearly upright by keeping your arms straight as you climb. Use the ball of the foot in climbing, and do not try to run up the ladder. Climb steadily and smoothly.

A leg lock on a ladder gives the brigade member a measure of security when both hands must be occupied with some manual task. A leg lock should be executed on the side of the ladder opposite that from which the work is to be done. Pass one leg between two rungs, locking the lower one behind the knee. Turn your leg so that the top of your foot is resting on the beam with the toes outside the beam and pointing downward. Now your hands are free to handle other tools.

Figure 5-23. Carrying short roof ladder up ladder, one man.

Figure 5-25. Placing longer roof ladder, two men.

Figure 5-24. Placing short roof ladder, one man.

Figure 5-26. Placing longer roof ladder, two men.

Figure 5-27. Climbing ladder.

Figure 5-28. Leg lock.

Figure 5-29. Dog chain hose tool.

Figure 5-30. Anchoring a ladder.

Other methods for securing yourself to a ladder include using a pompier belt or a rope tool. This sort of anchoring is generally done if one must remain in the same position for some time.

ANCHORING LADDERS

For safety's sake, it is important that a ladder be anchored to the building, especially during icy or windy weather. This can be done with a ladder "dog" chain, rope hose tool, or a hose chain or strap. Metal "dog" chains, which are passed around the ladder beams and spiked with a hammer or axe into the window sill, provide the most secure method of dogging a ladder.

Another method is to tie a rope to the ladder and to a bar across the inside of the window. The properly anchored ladder will not slip while it is being climbed or when the load is shifted.

Suggested Reading

NFPA 1931, *Standard on Fire Department Ground Ladders*, 1975, NFPA, Boston.

Fire Service Training Committee, *Handling Hose and Ladders*, 1969, NFPA, Boston.

Fire Service Ground Ladder Practices, 7th Edition, 1973, International Fire Service Training Association, Stillwater, Oklahoma.

Chapter 6

SALVAGE OPERATIONS

Salvage operations include all measures taken before, during, and after a fire or other emergency for the purpose of reducing loss from smoke, fire, and water and for minimizing business interruption.

An important part of the salvage job is preventing damage from water, whether from hose streams, automatic sprinklers, broken piping, floods, storms, or any other source. The water damage phase of salvage work involves protecting goods and machinery from contact with water, removing water from the premises with a minimum of damage, and limiting the amount of water used at fires to that actually needed for control and extinguishment.

Salvage involves the protection of property from the effects of heat, smoke, and contaminants in gases. This includes the removal of smoke and gases in the most efficient manner possible, using blower equipment where needed, and minimizing smoke odors with chemical deodorants immediately after the fire is extinguished.

Figure 6-1. Water in excess of that needed for fire extinguishment may sometimes have to be applied to ensure control of a fire. Craftsmanlike fire fighting includes salvage practices aimed at reducing secondary damage due to water, smoke, weather, pilferage, or other hazards.

Salvage includes the protection of records and equipment essential to the business during any fire emergency.

Salvage also includes restoration of the property to an efficient operating condition as quickly as possible after an emergency. This involves closing any roof, window, or floor openings that were made during fire fighting operations to gain access or to remove smoke or water. It involves the prompt servicing and lubrication of equipment, the restoration of heat or ventilation where required, and the restoration of utility services.

Salvage may also include the removal of debris from the fire or emergency area and the maintenance of plant security to protect the property by posting guards or locking the areas involved.

PLANNING FOR SALVAGE WORK

While fire brigade members may not have much to say about how plant manufacturing and storage areas are laid out and operated, much can be done in advance of a fire by plant management and department supervisors to reduce possible water damage.

There are three general measures commonly taken in an industrial plant to increase the effectiveness of salvaging operations, which may be performed by the fire brigade:

(*a*) Waterproofing of floors,
(*b*) Installation of scuppers and floor drains, and
(*c*) Protection of stock.

Waterproofing of Floors

Large fires requiring the use of heavy streams are likely to produce sizable water losses, which can be avoided to some extent if floors are reasonably tight. The extent to which floors will be made watertight will depend, of course, on the value and importance of the stock and equipment subject to possible damage. Completely watertight floors are difficult to provide. Ordinary concrete is not watertight, but it can be made reasonably so if cracks are filled with an asphalt mastic or are covered with an asphalt magnesite composition or a fine cement overlay.

Plank floors generally can be expected to leak even though a top flooring is provided. The same is true of floors of laminated construction. Some slight protection against leakage can be obtained in plank floors by the use of felt or building paper between the planking and the flooring. For installation over existing wooden floors, a hot asphalt mastic flooring is recommended for general industrial use. Other waterproofing arrangements include an asphalt emulsion flooring, a protective membrane flooring, and a magnesite flooring. In general, installation of these floorings requires considerable skill and attention to details.

Scuppers and Floor Drains

Waterproofing and drainage should permit the extinguishment of any fire without allowing leakage through the floor sufficient to cause material damage below. Waterproofing applied to floors should be carried up 4 to 6 inches (10 to 15 centimeters) at the walls and also around pipe shafts and other floor penetrations. Usually the most satisfactory type of drainage is to provide scuppers at the walls and, in some cases, additional floor drains connected to ample size leaders. The number of scuppers to be provided can be roughly calculated from an estimate of the number of hose streams likely to be used. For each hose stream of 250 gallons per minute (946 liters per minute), figure about two scuppers 4 by 4 inches (10 by 10 centimeters) or two floor drains connected to a 4-inch (10-centimeter) drainpipe.

Raised sills, ramps, or troughs should be provided at doorways or fire cutoffs. Drains should be kept clear and their locations posted by signs hung from the ceiling. This is particularly necessary in basements.

Protection of Stock

Protection of stock starts at the time it is piled. Workmen should understand that no pile of stock

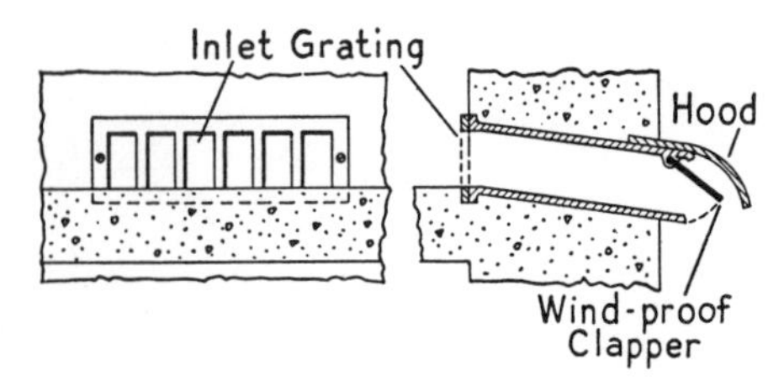

Figure 6-3. Floor elevation and section sketches of a type of scupper used for building drainage through an exterior wall.

should be so high as to be inaccessible for fire fighting. This means that no pile should be less than 3 feet (0.9 meter) from the ceiling and that space should be provided between walls and piles. A good rule is to keep stock in piles so that each pile can be conveniently protected by one standard salvage cover, size 12 by 18 feet (3.6 by 5.5 meters). Absorbent fibers and rolled papers require space to provide for expansion due to absorption of water. Improperly constructed shelves and wall cases may cause large water losses. Shelves should be constructed with a backing of wire netting about 2 inches (5 centimeters) from the wall, allowing a space of 18 inches (45.7 centimeters) at the top. Water following the wall will then pass without damage to the floor. In other cases, the top of the shelving may be covered with light metal attached to the wall extending about 6 inches (15 centimeters) in front of the shelves.

Placing stock on skids or pallets is always important. A general rule is that stock should be on skids holding the material 5 inches (13 centimeters) off the floor. This gives a chance for floor drains to work.

SALVAGE EQUIPMENT CARRIED BY THE FIRE BRIGADE

The equipment provided for salvage work varies and depends almost entirely upon local conditions. Equipment may be provided at convenient locations in the plant or carried on an in-plant emergency vehicle. Common tools and implements for salvage operations

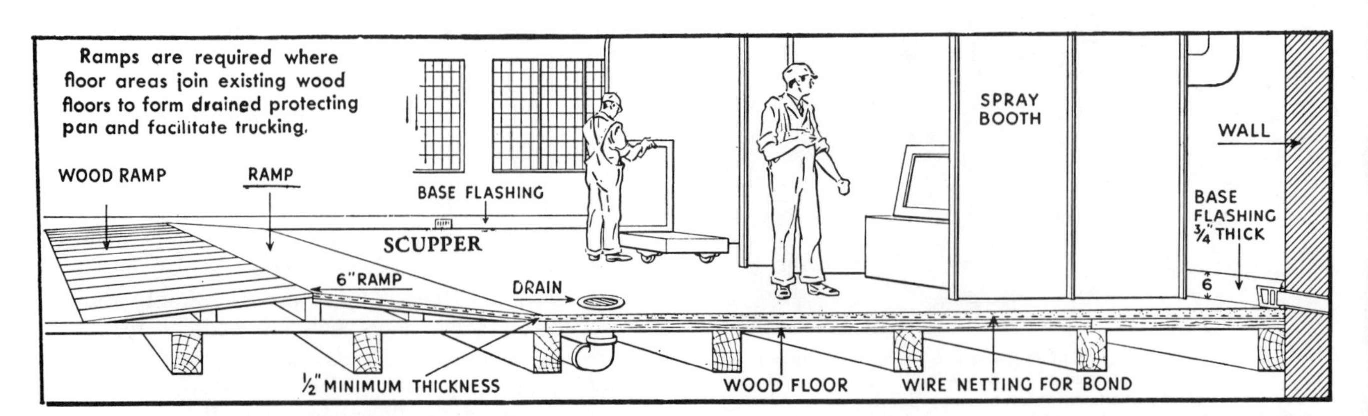

Figure 6-2. Paint spray operations are one of those found in industrial plants in which attention to water drainage is important. The sketch shows typical features, including asphalt mastic or oxychloride (magnesite) floor overlay, flashing, ramps, drains, and scuppers.

Figure 6-4. Pallets protect stock against damage from water on the floor.

in the area of which brigades should be trained are discussed in the following paragraphs.

Sprinkler stops are for temporarily controlling the flow of water from an open sprinkler until the system can be turned off, after which new sprinklers should be installed and the water turned on again immediately.

The common carpenter's claw hammer and an assortment of common and roofing nails are useful in salvage operations. They may be used to fasten salvage covers to walls and other objects where they would not stay in place without being fastened. A supply of roofing nails is necessary for securing roll roofing over holes through roofs, and other places where roofing can be used to protect the contents of a building exposed to the weather.

Heavy pressure-sensitive plastic tape in rolls, approximately 2 inches (5 centimeters) wide, may be used to fasten plastic sheeting or waterproof paper over window frames.

In the removal of water and debris from floors and in cleaning up and overhauling after a fire, scoop shovels are used.

Water can be swept with squeegees or ordinary brooms from floors into scoops and thence into containers and carried from the building. Vacuum cleaners for removing water are available. Wire brooms are useful in removing the heavier debris, such as dead embers and fallen plaster.

During cleanup after fires, merchandise and furniture should be carefully wiped and dried to remove water. Machinery must be dried and oiled. For those purposes, towels and wiping rags should be provided.

At least one roll of roofing should be carried to cover holes in roofs caused by fire or cut for ventilation purposes. Roofing paper serves this purpose fully as well as, if not better than, salvage covers.

Buckets and tubs are used for catching and carrying water and debris out of the building.

Drain screens or pieces of burlap are used to keep debris from clogging drain openings or openings cut in floors for drainage.

Following is a check list of equipment and tools useful in salvage operations. The number of each item to be provided is suggested, but it should be varied according to local requirements. The plant fire brigade should have certain items of such equipment and should know where other items may be obtained.

Principal items of salvage equipment:

Ten or more salvage covers, 12 by 18 feet (3.6 by 5.5 meters)
Two floor runners, 3 by 18 feet (0.9 by 5.5 meters)
Four brooms
Six 18-inch (45.7-centimeter) squeegees
Three scoop shovels
Two 10-foot (3-meter) ladders with folding roof hooks
Three claw hammers
One box assorted nails
One heavy-duty stapler
Six bale hooks
One smoke ejector, 5,000 cubic feet per minute (2.4 cubic meters per second) minimum capacity [with 100 feet (30 meters) of heavy-duty extension cord if electric]
One smoke deodorizer unit
Two rolls waterproof paper or plastic sheeting

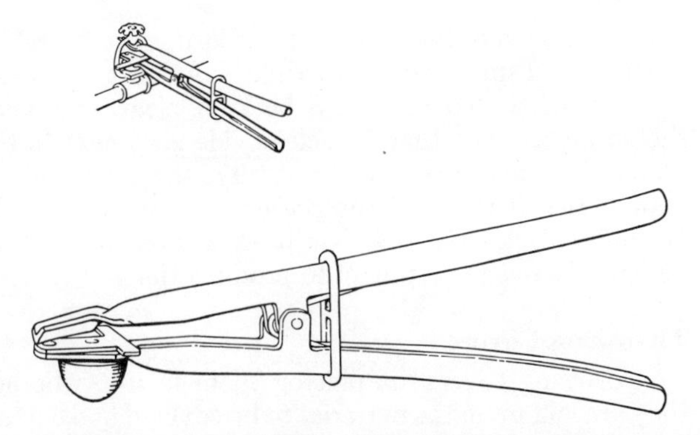

Figure 6-5. One variety of sprinkler stop or tongs.

Figure 6-6. Scoop shovel used in removing water and debris.

(Chicago Fire Department Photo)

Figure 6-7. Removing water from floors with squeegees.

(Scott Aviation, Lancaster, N.Y.)

Figure 6-8. Heavy duty vacuum cleaner for removing water.

One roll heavy-duty pressure-sensitive plastic tape, 2-inch (50-millimeter) width
125 feet (38 meters) Manila rope, ½-inch (13-millimeter) diameter
125 feet (38 meters) Manila rope, ⅜-inch (10-millimeter) diameter
Two padlocks and hasps, with keys
One water vacuum unit or low-head portable pump, electric driven with 100 feet (30 meters) of heavy-duty extension cord, or both, for dewatering

Forcible entry equipment:

Two axes, 6 pounds (2.7 kilograms)
Two hatchets
One crowbar, 36-inch (91-centimeter)
Two pinch bars
Two sledgehammers
One pick mall
Two cross-cut hand saws
One power saw [with 100 feet (30 meters) of heavy-duty extension cord if electric]
Two 8-foot (2.4-meter) pike poles

Sprinkler equipment:

Six sprinkler wedges or stoppers
One set sprinkler wrenches
One box assorted pipe plugs and caps
Three Stillson wrenches, sizes 10, 14 and 16 inches (25, 36, and 41 millimeters)
Two monkey wrenches, sizes 10 and 14 inches (25 and 36 millimeters)
Twelve or more sprinklers (assorted types and temperature ratings)

Electrical lighting equipment:

Two electric hand lights, 6-volt minimum
Three portable floodlights, 500 watt, including three 100-foot (30-meter) lengths of cable
One portable electric generator of at least 2,500-watt rating

USING SALVAGE COVERS

Any good grade of duck cover or tarpaulin is satisfactory for use as a salvage cover, provided that it is pliable and waterproof. The size of cover varies for different operations, but that most commonly used is 12 by 18 feet (3.6 by 5.5 meters), with grommets on all four sides.

Salvage covers should be folded and stored near where they are likely to be used or carried in a convenient place on a truck.

Folding and Carrying Covers

The method of folding a salvage cover is determined by the use to be made of the cover and the number of people available to use it.

There are many good ways of folding salvage covers. A fold designated as the "two-man" fold, not from the number of people who fold it, but from the number of people who use it, is illustrated.

Folding covers takes practice. Care must be taken to shake and smooth out all wrinkles. As far as possible, one side of the cover should be kept clean and folds should be so made that this clean side goes next to the materials being covered. Generally, the seam side is kept as the clean side. The procedure indicated should be followed closely, as it has been worked out by experienced salvage personnel to produce the best results.

Throwing Covers

"Throwing covers" is placing them in use, whether they are laid upon the material to be covered and spread out or ballooned over the material.

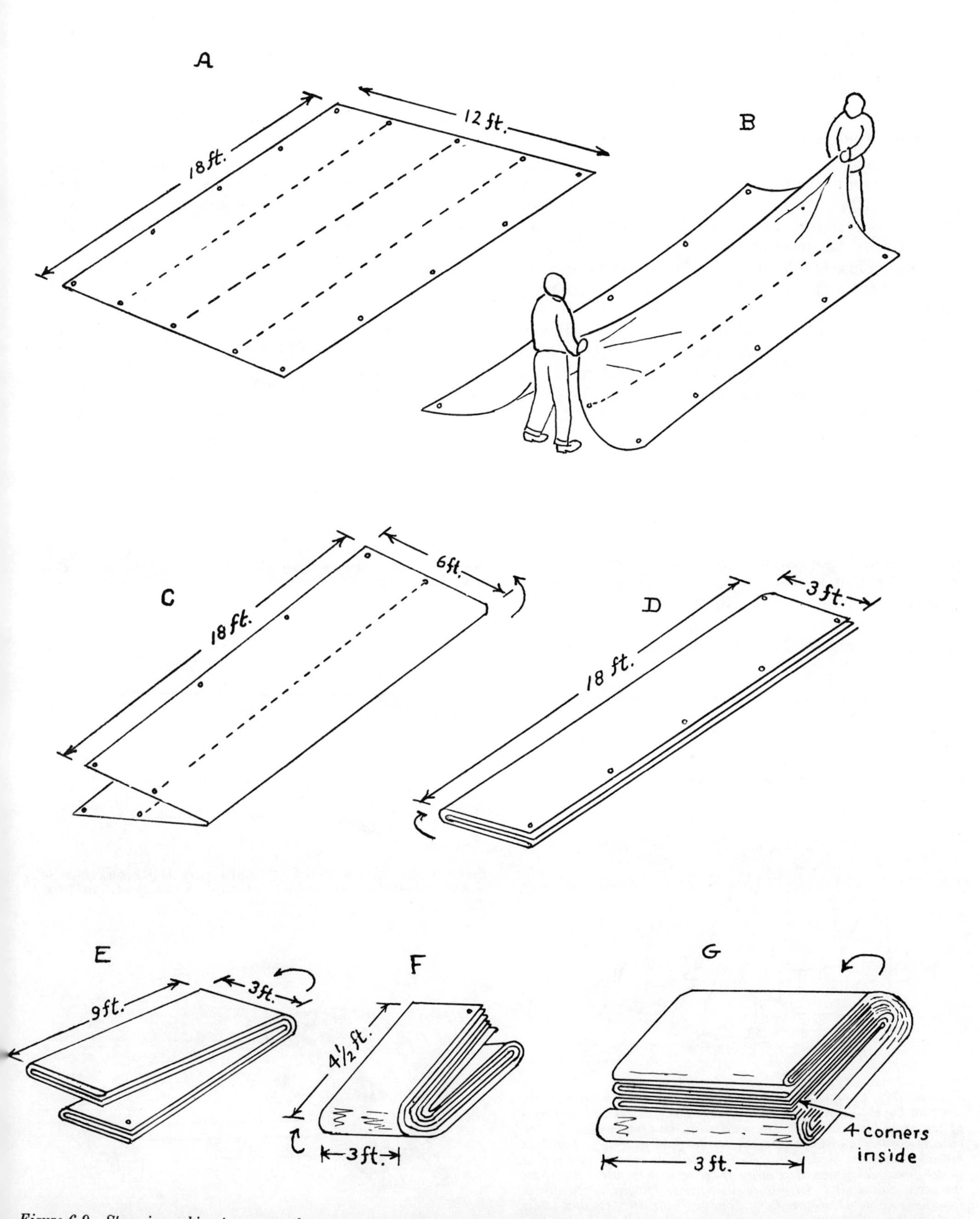

Figure 6-9. Steps in making two-man salvage cover fold. Start with cover flat (A). At each end, pick up the cover at the center fold (B), and, with a sweep, fold it once (C). Carry the single edges back in a second fold (D). Successive folds are illustrated (E, F, and G). A variation of the fold is to roll the cover after Fold E, and secure the roll with a string and slip knot.

The "two-man" fold described requires two people for throwing. The method of throwing the cover is shown. Practice of this procedure makes it possible for a crew to cover materials safely. Covers can be gently ballooned over fragile merchandise or manufactured articles on a bench or table.

Covering Shelving

In covering shelving, it may be necessary to drop the cover in a neat pile, and then carry it up and over the top. After placing the cover, tuck in the edges and ends. Exercise care in covering shelving. Many times it may be necessary to climb short ladders or to climb on insecure footing in order to place a salvage cover. Shelves and wall cases often are not properly built. Frequently they are fastened directly to the wall and ceiling, making salvage operations practically impossible. In this sort of situation, it may be more practical to remove the more valuable material from the shelves and cover it elsewhere.

REMOVING WATER

One of the best ways of limiting damage is to remove water from the building promptly. Where there is more water than can be removed from floors with squeegees and shovels, it is necessary to form catchalls and chutes that lead water out of windows and down stairways and to cut holes in the floor. Chutes can be formed with salvage covers and ladders. A chute thus formed will catch water coming from the floor or space above and carry it out a window. The same type of chute can be run down a stairway and out a door. If not enough covers are available to run to the door, sawdust may be used to form a trough.

Figure 6-10. One method of throwing the cover is to start with the cover on the shoulder of one fire fighter. The four corners and ends of center fold are inside. The man with the cover grasps the two outside corners and the end of the center fold nearest to him. A second man takes hold of the other two outside corners and end of center fold. The men step away from each other until they have the cover pulled out. In this position, it is hanging vertically in four folds (Fold D, Figure 6-9). The two outside edges and center fold are up. The throw operation is started with the two men thus holding the cover.

OPERATIONS IN SPRINKLERED BUILDINGS

When fire occurs in a sprinklered building, the operating sprinklers should be located as soon as possible by the company or brigade officer and salvage work started directly below.

Figure 6-11. The steps in throwing a cover can be demonstrated or practiced with an automobile as the object to be covered. The stretched cover (from Figure 6-10) is in two folds (four leaves). The fold nearest the car is dropped, one side being kept off the floor. This is so that the side of the cover next to the protected object is clean.

Figure 6-12. Second step in throwing cover is for each man to gather a handful of cover on the side toward which they plan to throw.

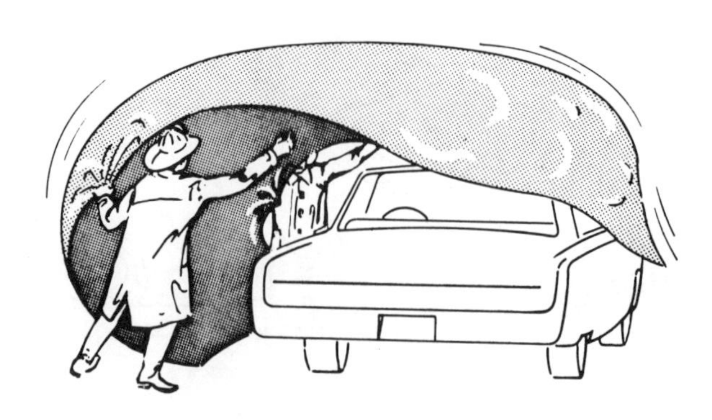

Figure 6-13. The throw balloons the cover on the air trapped under it.

(R. H. Redick, Fire Dept., Skokie, Ill.)

Figure 6-14. Electronic computer room and covers spread to prevent exposure to water and other hazards.

(Memphis Fire Department)

Figure 6-15. A good example of salvage cover work to protect equipment from water and falling plaster.

(Worcester Protective Department)

Figure 6-16. Shelving and goods piled on tables and chairs protected by salvage covers.

Sprinkler equipment should be turned off only upon orders of the officer in charge, and then only after it has been definitely determined that the fire is out. Leave a guard at the sprinkler valve until operation of the system is restored.

The sprinkler stop or sprinkler tongs are sometimes used to temporarily stop the flow from an open sprinkler. The system should be restored to service as promptly as possible.

OVERHAULING DAMAGED PROPERTY

Overhaul practices consist of certain operations conducted after the fire is believed to have been extinguished. The purpose of overhaul operations is to make sure that no hidden fire remains in walls, ceilings, stock, or other places, which might result in a rekindle after fire fighters have left the scene. There are numerous cases on record in which even a small amount of fire or embers left in the debris has caused reignition several hours after the initial fire. In some cases, the second fire has been worse than the first. In cases where there has been any appreciable amount of fire, it is customary to leave a so-called "watch line" and a detail. One or two fire fighters equipped with a hose line, an axe, and a pump tank or extinguisher should stay at the scene after other fire fighters have been released.

One of the most important aspects of the overhauling job is the proper handling of stock containing smoldering fire, such as bales of cotton, wool, or waste. Such stock should be moved outside the building as soon as the fire has been sufficiently controlled. Then the bales can be opened while a charged line is at hand from which water is applied sparingly to quench any embers.

Where only a small amount of fire is present it is often sufficient to use water pails in overhauling operations. A wetting agent applied with water used in this way reduces the total water required in some cases, keeping water damage low.

One of the first overhaul jobs is to remove water and water-soaked debris from the building to prevent further damage to floors, walls, and building contents. This should often not be done until authorized by the fire marshal or official fire investigator. All too often evidence of a fire started by an arsonist has been shoveled out before a fire investigator arrived. Scoops, brooms, and squeegees are used for this purpose. Particular care must be taken to watch for, and salvage, burned articles, including records or papers. They should be preserved, even though their value is seemingly small.

Holes made in roofs may be easily covered by inserting roofing paper under the remaining roof covering and fastening it in place. Pieces of old covers even one yard square are valuable for covering roof openings.

Holes in floors, broken windows, and the like may be repaired with boards, plywood, plastic sheets, waterproof paper, and other materials. If a door has been opened with care it may be closed and locked. If the fastening has been destroyed, a hasp and padlock may be supplied.

Another important part of the salvage and overhaul job is the prompt cleaning and oiling of machinery. In freezing weather the prompt restoration of heat is also of vital importance. Portable heating equipment used for removing dampness from buildings after a fire should be a safe type, properly installed.

Suggested Reading

NFPA 604, *Recommended Practice on Salvaging Operations*, 1964, NFPA, Boston.

The Fire Fighter's Responsibility in Arson Detection, 1971, NFPA, Boston.

Salvage and Overhaul Practices, 5th Edition, 1969, International Fire Service Training Association, Stillwater, Oklahoma.

Chapter 7

RESCUE WORK and EMERGENCY EQUIPMENT

In any fire fighting situation, there is always a possibility of someone being injured or unable to reach safety unaided. The primary function of the fire brigade is to control or, if possible, extinguish the fire quickly. However, it may be called upon to aid trapped and injured coworkers simply because there is no time to await the arrival of trained emergency medical teams and the fire department rescue crew.

Fire brigade members should not be expected to do the work of emergency medical technicians, but they should know something about basic first aid and rescue operations. If the victim is immobilized and in no immediate danger of further injury, he or she should not be moved until trained help arrives. Where necessary, assist respiration or breathing and control severe bleeding. Make the patient as comfortable as possible and try to reduce shock.

These actions cover most fire rescue or emergency problems that may be encountered by industrial fire brigade members.

A first-aid course is not included in this text. It is desirable that each member of an industrial fire brigade complete at least a basic first-aid course, such as is offered by the American National Red Cross and other agencies. The necessary classes can be arranged locally. A number of the members of any brigade should qualify as instructors in first-aid work.

COMMON EMERGENCIES

Some of the leading reasons for rescue calls received by fire brigades include persons trapped due to fire, smoke, gases, or lack of oxygen. The victims may be trapped in burning buildings, in vats, manholes, or excavations. Persons may be trapped or caught in elevators, hoists, lifts, or machinery, or may be pinned under machinery or vehicles. Persons not trapped may be suffering from injuries, such as cuts, burns, broken bones, interruption to breathing, poisoning by various chemicals or gases, that may prove serious or fatal unless effective first aid is given promptly.

Bleeding

Humans can live only a short time if severe arterial bleeding (indicated by spurting of blood) is permitted to go unchecked. Familiarity with the few pressure points that will permit control of bleeding from the various arteries has saved many lives. Cuts due to broken glass and other sharp objects are a frequent occurrence at fires and explosions. It is helpful to have

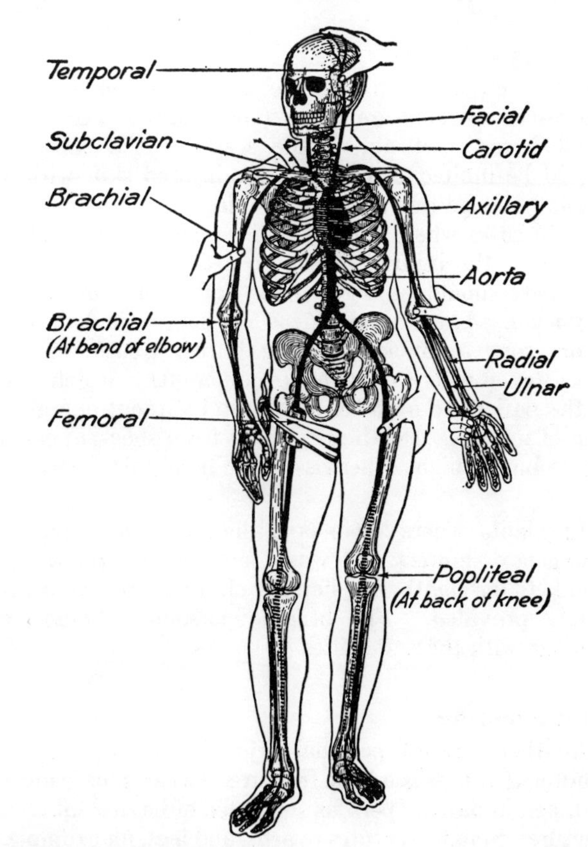

(U. S. Bureau of Mines)
(Manual of First Aid Instruction)

Figure 7-1. Points at which pressure must be applied to control severe arterial bleeding.

a chart of these pressure points posted as a reminder and to practice control of bleeding in first-aid and rescue classes.

Interrupted Breathing

Once breathing has stopped, a person can live only a very few minutes. Interruption to breathing can be occasioned by smoke, gases, electric shock, immersion, poisons, and other causes. The rescue worker must be familiar with means of resuscitation of persons suffering from lack of oxygen or other respiratory difficulties. In many cases, excellent mechanical devices including inhalators and resuscitators are available to assist respiration, but it is of vital importance that manual methods of assisting respiration (mouth-to-mouth, for example) be started at once as soon as the rescue worker reaches the patient or brings him to a respirable atmosphere. A minute or two lost while awaiting the arrival of mechanical devices may prove fatal.

Brigade members should learn one or more of the approved methods of artificial respiration. Circumstances, such as personal injuries and location of a trapped victim, may not always permit the use of one method; therefore, rescue workers should be familiar with several other effective manual methods of assisting respiration.

Burns (and Scalds)

Other important injuries frequently encountered by fire rescue workers are burns (and scalds). For most small burns, first aid at the scene of the accident should be limited to covering the injured skin with a sterile bandage only. There are dressings that can be used in cases where there would be considerable delay in getting the injured person to a doctor or hospital, but these should be selected for fire brigade use by a physician. The first-aid measures should include extreme care to avoid infecting burns, treatment for shock (even where shock is not apparent) and delivery of the patient to a doctor or hospital without aggravation of injuries. Death may result from shock in cases where burns might otherwise prove nonfatal.

In plants where there is a danger of acid burns, emergency showers and wash facilities are provided. Blankets for smothering flaming clothing are also commonly provided. Fire brigade personnel should be familiar with these facilities.

Bone Fractures

Another type of personal injury that may be encountered at fires is a bone fracture. Greatest care must be taken in moving persons suspected of having suffered fractures. Simple fractures to arms and legs, for example, may be made compound (bone pushed through the skin) by careless movement. Paralysis or death may result from careless movement of persons suffering from fracture of spinal vertebra. Proper first-aid procedures must be followed to immobilize or make rigid any suspected broken bones if it is necessary to move a patient suffering from fractures. One advantage of requiring all members of a fire brigade to take first-aid instructions is that they learn how to move persons with a minimum chance of aggravating injuries.

PLANNING FOR EMERGENCIES

If it is to deal successfully with emergencies, the fire brigade must do some preincident planning. The brigade as a whole should be aware of its responsibilities, and the individual members should be schooled in their specific assignments.

By virtue of their normal duties in the plant, some brigade members may be better suited for certain assignments than others. For example, production department personnel might be best qualified to shut down a critical manufacturing process that may be endangered by, or contribute to, an emergency situation. Similarly, maintenance department personnel may be the logical people to operate fire pumps and replace sprinklers. Others may be assigned to fire fighting duties.

In making its plans, the brigade should divide the plant into sections and consider what type of emergency might be expected to occur in each. Such factors as the nature of the hazard in a given area, emergency equipment available, and each brigade member's normal work location in relation to the emergency site will have to be taken into account when personnel assignments are made.

It is important that the fire brigade know what its relationship to the public fire service is to be, so that there is no confusion when the fire department arrives on the scene. Toward this end, it is highly desirable that the public fire department participate in planning and training sessions with the fire brigade.

Planning for fire emergencies should also include emphasis on safety in the training of fire fighters. In all of the jobs taught in this manual, including the proper use of fire fighting tools such as hose, ladders, and axes, safety is stressed.

PROTECTIVE CLOTHING

An essential element in both fire fighting and rescue work is the wearing of proper protective clothing. Fire fighters' coats, boots, and helmets are intended not so much for protection against water (fire fighters expect to get wet when fighting a fire), but for protection against extremes of cold and heat, burning embers, cuts and scratches from falling debris.

Fire fighters' helmets not only must withstand hot water and embers, but also must protect the head against falling objects and objects that may be bumped against in the darkness or smoke. Fire fighters' boots should have an innersole to protect against nail punctures — a frequent source of accidents at fires. A good pair of work gloves that can stand water are valuable protection to the hands.

Often it is impossible to get into areas where it is necessary to make rescues unless protective clothing is worn. The fire fighter's ability to work in smoke and heat comes through training, experience, and the use of protective clothing.

RESCUE TOOLS AND EQUIPMENT

Most industrial plants have many of the tools and equipment that are needed in rescue work. The people with special skills to use them are often selected as members of the fire brigade or, in large plants, as members of special rescue crews. These include individuals who can use cutting torches, riggers familiar with rope tackle, electricians, pipe fitters, and other workers.

The fire brigade or rescue squad should check the availability of such equipment not mentioned elsewhere in this chapter, but listed below, and provide items it does not have, but which are desirable for the particular plant.

- Blankets.
- First-aid kits.
- Litters or stretchers.
- Pompier belts.
- Jacks; 5-, 10-, and 20-ton (4,500-, 9,000-, and 18,000-kilogram).
- Acetylene cutting outfits.
- Boats for water rescue.
- Suits with hoods for ammonia wading.

RESPIRATORY PROTECTIVE EQUIPMENT

A valuable tool in many fire and rescue situations is respiratory protective equipment. It allows the fire fighter or rescue worker to operate in an atmosphere that one normally could not tolerate for any appreciable length of time or indeed at all. Only units meeting the requirements of the National Institute of Occupational Safety and Health (NIOSH) should be made available for industrial fire brigade use.

One of the first things to be determined when a rescue must be made is whether the victim is in a breathable atmosphere. If the person is unconscious in a below-ground location, such as an excavation, manhole, basement, vat, or tank, the likelihood of oxygen deficiency or an accumulation of toxic gases must be considered. Many rescue workers have died alongside of the victims they had hoped to save, because they plunged into an unbreathable atmosphere without stopping to don breathing equipment. In all such situations, suitable breathing equipment is essential and must be promptly available.

In such rescue work, rope hitches for body ties are used. The rescuer must wear a lifeline when making this type of rescue.

Ventilation fans or blowers, such as are employed for smoke extraction, may be useful in rescue operations in dangerous atmospheres. However, care must be taken to guard against ignition of any flammable concentration of gas or vapors.

When working in smoke, move deliberately whether you are wearing breathing apparatus or not. Unnecessary activity wastes energy and causes heavy breathing. Without a mask, the heavier you breathe, the more smoke and heat you will inhale. With a mask, the heavier you breathe, the quicker you deplete your supply of air.

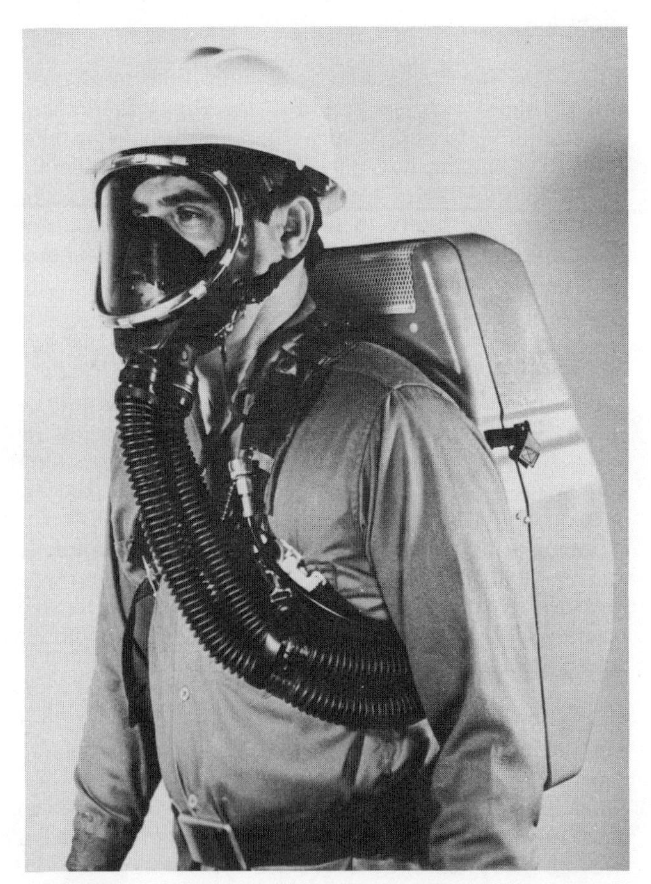

(Scott Aviation, Lancaster, N.Y.)

Figure 7-2. A Scott Rescue Pak is a four-hour closed circuit breathing apparatus that contains its own oxygen supply.

There are several types of respiratory protective equipment. Each type is designed for a particular type of service and must be used within the limits of the protection it will provide. NIOSH lists self-contained breathing apparatus and other protective equipment, describing the conditions for which each is approved. Limitations set forth by NIOSH and in the manufacturer's instructions should be closely observed. The protective devices being considered here should not be confused with simple respirators or filters used in connection with slight dust and fume conditions.

Regardless of the type of respiratory equipment that is available in your plant, you should be thoroughly trained in its use before wearing it under emergency conditions. Never work alone when using protective breathing apparatus; always have at least one other person working with you.

Canister-type Masks

Canister-type gas masks consist of a tight-fitting facepiece connected to a canister containing chemicals that remove parts of gases and irritating substances from otherwise breathable air. Special canisters are available for industrial use to provide protection against specific gases, such as ammonia and chlorine, but they are not intended for use in fire fighting operations.

Canister masks have serious limitations and ordinarily should not be provided for fire brigade use. They are able to absorb only small amounts of noxious gases from the air, and they do not contain or generate the oxygen necessary for life support. Toxic gas concentrations are likely to be high and oxygen concentration is likely to be low during combustion. Therefore, canister masks should not be used in fire fighting operations.

Self-contained Breathing Apparatus

Where fire fighting and rescue operations must be performed in an atmosphere deficient in oxygen or laden with toxic gases, wear self-contained breathing apparatus. Self-contained apparatus provides the air or oxygen necessary for survival. There are three types of self-contained breathing apparatus available for fire fighting applications. Regardless of the type provided in your plant, you should have had extensive training and practice in its use before wearing it under actual emergency conditions.

One type is a closed-circuit system equipped with a canister and a breathing bag. This apparatus chemically generates oxygen after the canister seal has been punctured and the wearer has exhaled into the mask to start the reaction. It is called a closed-circuit system because the wearer's breath is not released to the atmosphere, but is circulated back to the canister to continue the oxygen-generating reaction. The length of time for which each canister will provide protection depends, in large measure, upon the activity and oxygen consumption of the wearer. The canister is good for only one usage after the seal has been broken; therefore, the wearer should be responsible for installing a new canister after each use.

("Chemox" breathing apparatus
Mine Safety Appliances Co., Pittsburgh, Pa.)

Figure 7-3. Self-generating-type breathing apparatus that utilizes a chemical canister, which evolves oxygen and removes exhaled carbon dioxide in accordance with individual breathing requirements. The retention of moisture from the exhaled breath permits fog-free lenses. The mask is approved by the U.S. Bureau of Mines for 45 minutes of service in oxygen deficient atmospheres. Each canister is good until the seal is broken, but a new canister must be installed after each use.

Another type of self-contained breathing apparatus uses a cylinder of oxygen, a breathing bag, and a container of chemicals. This type is a closed-circuit, oxygen rebreathing system. The wearer's exhaled breath passes through the container of chemicals, where carbon dioxide is removed. It is then mixed with oxygen in the breathing bag and inhaled again.

The third type is an open-circuit system using a tank of compressed air or oxygen, a pressure regulator, and a face mask. This type is called a demand mask. When the wearer inhales, the regulator valve opens, allowing air or oxygen to enter the mask. As the wearer exhales, the supply valve closes and the exhaled breath operates an exhaust valve and is vented to the atmosphere. After each use, the tank should be

("Air-Pak" breathing apparatus
Scott Aviation, Lancaster, N.Y.)

Figure 7-4. Self-contained breathing equipment supplying air through a demand valve and approved by the U.S. Bureau of Mines for 30 minutes in oxygen deficient atmospheres.

refilled or replaced with a full tank before the apparatus is returned to storage.

Hose Masks

Hose and fresh air masks are not commonly used for fire fighting, but have certain usefulness where masks are needed in some industrial operation. Such mask assemblies consist of a facepiece, air hose, and a container of air under pressure. Instead of a container of air, some assemblies use a blower to carry outside air through the hose to the face mask. An obvious limitation of this type of apparatus is that it is impractical for more than one person to be dependent on a single air line. The fresh air mask has the limitation that even the outside air near a fire is likely to be heavily contaminated with smoke. Such masks may be useful, however, where rescues are needed from oxygen deficient areas in tanks, vats, or manholes and where openings are not large enough to permit rescue workers wearing self-contained breathing apparatus to gain entrance.

(Mine Safety Appliances Co., Pittsburgh, Pa.)

Figure 7-5. Demand-type self-contained breathing apparatus rated for 15-minute service life.

(Demand type breathing apparatus
Mine Safety Appliances Co., Pittsburg, Pa.)

Figure 7-6. Demand-type self-contained breathing apparatus rated for 30-minute service life.

Using and Caring for Breathing Apparatus

In preparation for putting on the mask of breathing apparatus, grasp the facepiece in both hands, keeping thumbs inside the mask and below the lower head harness strap and keeping fingers outside the facepiece.

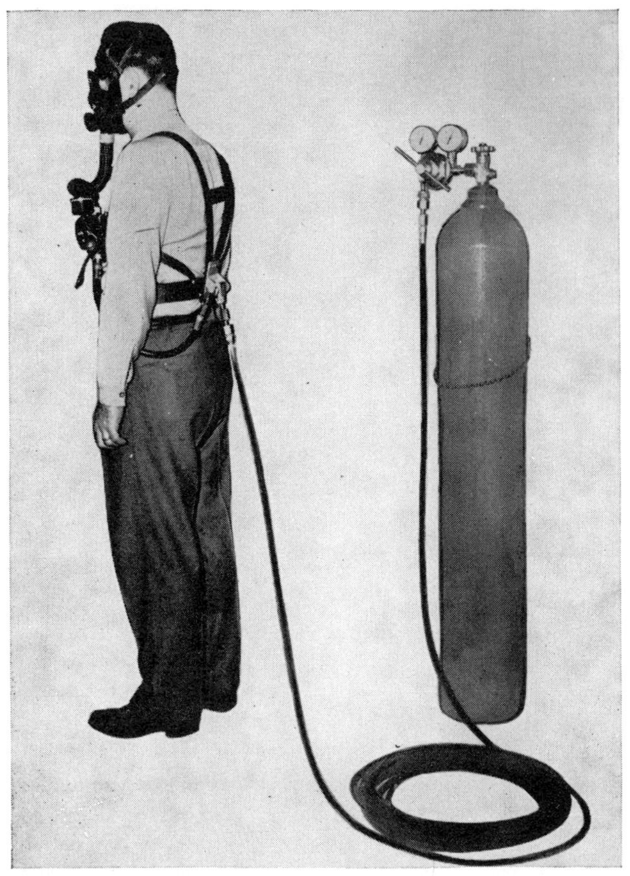

(Scott Aviation, Lancaster, N.Y.)

Figure 7-7. An air line mask designed for long duration work in oxygen deficient or lethal atmospheres. This mask has a demand regulator and is supplied by an air line from an air tank. A similar type is an air hose mask that employs a blower to carry outside air through a hose to a mask face piece.

Pull the mask apart with your thumbs, and raise it to your face. Thrust your chin upward and outward and fit it into the lower part of the mask. Press the mask against your face, and pull the head harness into position on the back of your head. Pull on the harness straps so that the mask is tight against your face. Cut off the air supply, and try to inhale. If the mask is airtight, it will collapse. If it is not airtight, readjust the harness straps.

Practice using the breathing apparatus to build the confidence necessary for working under restricted breathing conditions. Practice will also teach you the limitations of the equipment. It is obvious that you will not be able to work as freely when wearing breathing equipment as you would in an unlimited fresh air supply. It is also important that you begin using the equipment before becoming fatigued by previous heavy exertion in heat or smoke.

Make sure that the apparatus is clean before it is put away. Store it in a cool dry place. Handle breathing apparatus with care in appreciation of the fact that it may save your life. Before breathing apparatus is put away, be sure that it is ready for the next emergency. See that air or oxygen tanks are full or that new canisters have been installed.

Suggested Reading

Bahme, Charles W., *Fire Officer's Guide to Dangerous Chemicals*, 1972, NFPA, Boston.

Fire Officer's Guide to Breathing Apparatus for the Fire Service, 2nd Edition, 1974, NFPA, Boston.

NFPA 49, *Hazardous Chemicals Data*, 1975, NFPA, Boston.

Fire Service Rescue and Protective Breathing Practices, 4th Edition, 1969, International Fire Service Training Association, Stillwater, Oklahoma.

Fire Service First Aid Practices, 5th Edition, 1976, International Fire Service Training Association, Stillwater, Oklahoma.

Erven, Lawrence, *First Aid and Emergency Rescue*, 1970, Glencoe Press, Beverly Hills, California.

American Red Cross, *Standard First Aid and Personal Injury*, 1973, Doubleday & Company, Garden City, New York.

Chapter 8

GENERAL PRACTICE in FIRE FIGHTING

In each property, procedures for dealing with fires must be developed. Certain fire fighting procedures are basic. However, plans for each individual property may differ because there are different things to burn and different facilities for fire fighting. It is important to have operating procedures designed for each property. It is assumed that properties using this manual have developed their fire brigade organization and assigned its duties.

In some properties or in some areas, industrial processes are sufficiently complicated that only the operating superintendent is qualified to say what fire control measures should be taken. In such cases, the fire brigade's duty may be to operate special equipment or to provide hose streams at the direction of the operating superintendent. However, in most industrial and other properties, management leaves decisions of fire fighting procedure to the fire brigade chief or officer in charge. In some properties, there is a fire loss prevention manager who is often a fire protection engineer familiar with the various hazards and desirable fire control methods. The loss prevention manager should be consulted.

This text is confined to standard practices, which, if followed, will increase the chances of controlling fires with a minimum of damage and, therefore, shorter interruption to operations.

It is particularly important that management as well as officers of the fire brigade examine the procedures that have been established for fire control. Factors that should be subject to such prefire planning would include:

(*a*) Storage and handling methods for new materials and finished products with regard to likelihood of ignition and fire control measures that may be needed.

(*b*) Process hazards, particularly of flammable and explosive or toxic materials. This should include identification of hazards, special protective equipment needed, and fire fighting procedures.

(*c*) Measures for limiting the size of a possible fire. This includes some structural design features, such as fire walls and stair enclosures. It also includes provision of fire fighting equipment and tactical plans for confining such fires as may occur.

(*d*) Fire fighting equipment, including automatic sprinkler systems with adequate water supplies at required pressures, fire hose, nozzles, and other fire brigade equipment.

(*e*) Management instructions to department heads, supervisory personnel and general employees. These will include maintenance and fire prevention measures, procedures to be followed in case of fire in various departments, the duties of fire brigade members and fire guards, and procedures to be taken following fires to get the property back in operation with a minimum of disruption. The latter would also include proper investigation of factors responsible for the loss and preparation of reports.

(*f*) Evacuation plans. All employees should know what to do if it becomes necessary to evacuate all or part of the plant. A detailed plan is necessary to ensure safe, orderly plant evacuation.

For many properties, one fire that is not controlled immediately can be disastrous. Management and the fire brigade may not get a second chance if production is seriously interrupted by fire. Some of the most important fire control measures must be taken before fires occur. These include keeping equipment ready and training both brigade members and plant employees in emergency procedures. Fire fighting and evacuation plans that have been adequately rehearsed should go into effect smoothly when an actual fire occurs, despite the confusion that tends to accompany emergency situations. Fires should be fought three times — before the fire, for planning; during a fire, to carry out plans; and afterward, for the lessons the fire teaches.

GENERAL FIRE BRIGADE PROCEDURE

It is assumed that, before the fire, the major features of how a fire will be attacked will have been worked out. A well-trained brigade should have held practice sessions at which the brigade responded to simulated alarms in all major parts of the property. Exact details of what the brigade does depends on circumstances according to the property area affected.

In industrial fire protection, principal reliance is placed upon private fire protection equipment, which should be installed in accordance with the hazards to be protected. Fire extinguishing equipment may include (1) portable fire extinguishers of various types and sizes appropriate for the hazards, (2) automatic sprinkler systems, (3) standpipe and hand hose systems, (4) hydrant and hose stream systems where fire hose streams may be needed, and (5) special extinguishing systems, such as foam, carbon dioxide, dry chemical, and halogenated extinguishing agent systems as needed to cope with special hazards.

In a property with a well-trained loss prevention organization, fire fighting will begin with things done by the employees. These primary functions will be under the department heads and concern employees in each department assigned to fire fighting or related duties. Such employees are sometimes described as fire guards. They are not necessarily members of the plant fire brigade, but will usually be trained to use fire extinguishers, give the alarm signal, and perform other duties that have to be carried out in case of fire in their area or department of the property.

Usually, in industrial fires, the first attack on a fire will be made with portable fire extinguishers located adjacent to the fire area. Normally, they will be used by employees working in the area or by plant guards or patrolmen. In many cases, fires will be extinguished before automatic sprinklers operate or before the fire brigade crews arrive.

It is important that, when extinguishers are used, the fire brigade chief is notified promptly, even if the fire appears to be out. There are a number of reasons for this. Frequently, fire or sparks have extended to some concealed area or process, and the fire has not been extinguished. The fire brigade chief will need to see that used extinguishers are recharged promptly and that substitutes are provided while recharging takes place. The chief should check the automatic fire protection devices in the fire area to see that fusible links or sprinkler solder has not been weakened by heat and that salvage operations and clean-up are properly conducted. It is also important to make a record of each fire incident, however small. This is necessary in measuring the success of the property's fire loss prevention program, in determining the processes and hazards causing fires, as well as in determining not only the direct fire losses, but also the cost in labor and interruption of plant production. Although most fires will be extinguished quickly when they are small, serious losses are likely to occur from time to time unless the factors contributing to fires are corrected. Therefore, each fire must be considered as serious and also as a test of the property's fire loss prevention organization.

A prompt sounding of the alarm by employees in the fire area should get the fire brigade on its way. If departmental employees have the fire under control when the brigade arrives, the brigade members should first check for extension of fire. They should ventilate smoke and begin necessary salvage and clean-up work.

When an alarm of fire is received at the property's fire alarm communication center, established procedures should be followed to notify the public fire department. These will vary with the location of the property and the degree to which the brigade is organized and equipped to provide the required initial response to alarms of fire. In cities having paid fire departments, municipal fire officials take a dim view of a firm's failure to notify the public fire department until after a fire is beyond control of the plant forces. The procedure should call for immediately transmitting an alarm to the public fire department and notifying personnel at the property gates of the location of the alarm. If the fire is extinguished before the public fire department arrives, the officer in charge of the brigade can so advise the gate guard. The fire department officer may then merely make an inspection and obtain necessary data for the incident report. Where the fire has not been extinguished when the department arrives, the operations should be carried out in accordance with plans. One of the important functions of the fire department would be to promptly make connections to pump water into the automatic sprinkler system to back up the initial supply. Other routine fire fighting operations would be carried out as needed, including raising ladders to the roof if required for access to the fire or ventilation.

If a property is located in a community having a fire department with limited personnel or is located some distance from either public or private fire departments, the plant brigade is likely to be on its own during the critical early stages of a fire. It is poor practice to call for assistance after a fire has spread out of control. It takes time for outside help to arrive, and the plant fire brigade chief should be given the responsibility for calling adequate assistance as soon as it appears that a fire may tax the ability of his own force. Calling for assistance is high on the list of responsibilities of a chief officer whether of a plant brigade or a public fire department.

In some large properties, the fire brigade consists of full-time fire fighters. They may be organized into fire companies, which will respond to alarms from their station or stations just as public fire departments do. However, many fire brigades are composed of fire fighters on call who have other duties and who respond to a fire signal. Some of these members may be required to report to a central point in the property to man available equipment. However, where a system of coded fire signals is used, some members may be assigned to respond directly to the area of the plant concerned to man equipment already provided in that area. Obviously, members of the brigade must know the various signals to which they must respond.

USING STANDPIPE HOSE SYSTEMS

Standpipe hose systems are provided to furnish prompt hose stream service inside of buildings. While many fires are readily extinguished with portable extinguishers, most extinguishers have a limited period of discharge. Hose streams provide a continuous extinguishing capability, which may be applied as needed until the fire is out. Standpipe systems commonly are equipped with small hose — usually a 1½-inch (37-millimeter) diameter waterway — for use by building occupants. Some standpipe systems are provided with outlet valves for supplying 2½-inch (62-millimeter) fire hose for use by a trained fire brigade or the public fire department when standard hose streams are needed. Some systems have both large and small hose outlet valves.

Streams from standpipe systems may be valuable in conjunction with fires being held in check by automatic sprinklers. Some fires may burn in places out of direct reach of sprinklers, such as under work benches, in piled storage, or on shelving. Judicious use of streams from standpipe hose will permit extinguishment of such fires with a minimum of loss.

The simplest form of standpipe hose equipment consists of unlined hose with a straight stream nozzle and no nozzle shutoff. Normally, such equipment is found in locations where hose streams may be needed only on rare occasions but where there may be delay in running hose lines from the outside. Such locations include the upper floors of buildings.

Where fires are likely to be more frequent, as in industrial operations, standpipe systems employ lined fire hose equipped with adjustable spray nozzles. Lined hose is a more efficient conductor of water and does not permit the seepage of water through the hose jacket associated with the use of unlined hose. Therefore, better nozzle pressure is obtained from a given pressure at the standpipe, and water damage in the area may be less. The adjustable spray nozzle allows a selection of stream patterns from a wide spray useful on flammable liquids fires to a virtually straight stream, which may be employed where reach and penetration of stream are desired.

The training of fire brigade personnel and other plant employees who are expected to use standpipe hose systems must cover use of the hose provided.

Commonly, the hose is on racks or reels. Proper operation normally requires at least two brigade members. One should take the nozzle and advance the line toward the fire, being careful to straighten the line along the way. The other assists in removing hose from the rack and keeping the line reasonably straight. All the hose should be removed from the rack before the water is turned on. This is to prevent kinking and possible bursting of the hose. If the line kinks, water will be cut off until the kink is removed. Once the line is charged, the second fire fighter can advance to assist the person at the nozzle in fighting the fire. Proper manning of the line permits it to be mobile, so that maximum effectiveness may be obtained.

Where shutoff nozzles are not provided, it may be desirable to have a brigade member remain close to the valve to shut off the flow of water as soon as the stream is no longer needed. There is little advantage in continuing the direction of a stream at a location where the fire has already been extinguished.

Brigade members should practice the use of various spray patterns so as to get maximum effectiveness from the stream provided by an adjustable nozzle. Usually a straight stream should never be directed into a container of flammable liquids or at live electrical equipment.

Streams from 1½-inch (37-millimeter) standpipe hose may discharge from 25 to 50 gallons (95 to 190 liters) of water per minute. They are designed for pressures as low as 25 pounds per square inch (172 kiloPascals). By contrast, 1½-inch (37-millimeter) lines of hose supplied by fire department pumping engines will discharge between 60 and 125 gallons (227 and 473 liters) per minute at nozzle pressures of 80 to 100 pounds per square inch (552 to 689 kiloPascals).

USING STANDARD HOSE STREAMS

There are numerous fires in industrial plants and other private properties maintaining fire brigades, which require the application of standard streams from 2½-inch (62-millimeter) fire hose. A standard hose stream discharges approximately 250 gallons (946 liters) of water per minute. It is based on discharge from an "Underwriters" playpipe with a 1⅛-inch (28-millimeter) solid stream orifice at 45 pounds per square inch (310 kiloPascals) nozzle pressure. This pipe has no shutoff valve so the water must be turned on and off at a hydrant or other hose supply valve.

Adjustable spray nozzles capable of both spray and straight streams are preferable when using 2½-inch (62-millimeter) fire hose. Spray nozzles may deliver somewhat less water than a standard hose stream. This is because these nozzles are designed for use at somewhat higher pressure and the nozzle reaction may be too great for convenient handling unless the flow is somewhat reduced.

Among the places where standard hose streams may be needed are yard storage, loading platforms, roofs, high piled storage in buildings, unsprinklered buildings, exposure fires in adjoining properties, and large storage tanks that must be kept cool during a fire.

Standard hose streams are supplied from the following sources: yard hydrants, roof hydrant outlets, large standpipe outlets, and by automotive pumping apparatus where available. In most locations there will be a public fire department available on call to respond with pumping apparatus and hose.

The fire brigade training program should include running 2½-inch (62-millimeter) fire hose to the various places where standard hose streams may be needed should fires occur. This is a very important exercise. Practice is needed to get nozzles quickly to the desired points of application. Also, this will make the brigade members familiar with the amount of hose needed to reach various locations. It is considered very poor fire fighting to "lay out short," that is, not having enough hose to direct the stream at the place where it is needed.

Normally, it is good practice to have at least 50 feet (15 meters) of hose in the line near the point where the nozzle is to be operated. This permits the stream to be moved about for most advantageous operation. However, in using a nozzle on a large hose line, the hose should be relatively straight for about 20 to 25 feet (6 to 7.6 meters) back of the pipe. This reduces the tendency of the hose to try to whip and straighten itself. If a hose whips, the nozzle may get away from the fire fighters at the pipe, and this can cause injury. It is generally good practice to operate lines at a conservative pressure that will permit some mobility or flexibility in the direction of the stream. Pressure at the nozzle can be increased on orders from the person in charge of the stream when it is safe and desirable to do so. If the pressure at a nozzle is too great for the available brigade members to handle, the line should be lashed to a support before water is turned on. The nozzle reaction also can be reduced by shutting the valve partway until the pressure can be reduced at the source.

Where hose houses are distributed about a plant yard, usually there is a relatively short hose line having a nozzle connected to the hydrant. This may be used where the fire is close to the hydrant.

Normally, additional hose is stored in the hose house. Where the fire is farther away from the hydrant or where it is desired to bring additional streams into play, it may be necessary to piece out hose lines to permit longer stretches. Brigade members should practice extending hose lines as quickly as possible. Later chapters will deal with general fire fighting tactics and development of effective hose streams.

FLAMMABLE LIQUID FIRES

Some of the most troublesome types of fires encountered by industrial fire brigades are those involving flammable liquids. These include fires in various manufacturing and finishing processes where flammable liquids are employed as well as fires involving stored flammable liquids. Flammable liquids are usually in containers where a blanketing or smothering effect is effective for extinguishment. By contrast, ordinary combustibles are controlled by extinguishing equipment using cooling and quenching action.

Due to the often more spectacular burning of flammable liquids, as compared with ordinary combustibles, and the tendency of flammable liquid vapors to flash or explode under certain conditions, fire fighters have sometimes been discouraged from making adequate attempts to control such fires. Actually, however, if the flammable liquids in a given plant are being used in accordance with recognized good practices and proper

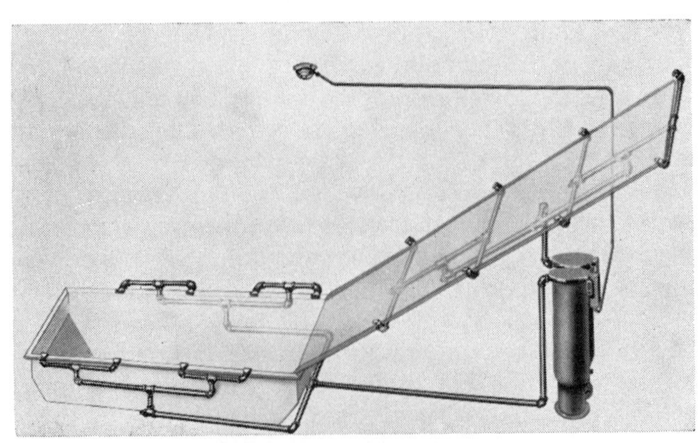

(American-LaFrance-Foamite Corp., Elmira, N.Y.)

Figure 8-1. System for automatically applying foam to paint dip tank and drainboard.

(American-LaFrance-Foamite Corp., Elmira, N.Y.)

Figure 8-2. A foam protection system for storage tanks includes piping, hydrants, and outlets.

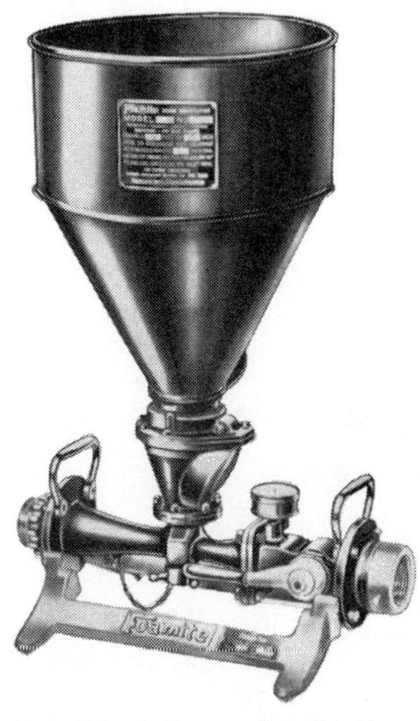

(National Foam Systems, Inc., West Chester, Pa.) (American-LaFrance-Foamite Corp. Elmira, N.Y.)

Figure 8-3. Foam generators for installation in hose or pipe lines. In the type illustrated, the foam-making ingredients are premixed and stored in cans ready to be dumped into the hopper. The generator mixes the powder with the water passing through the line. Generators with dual hoppers and line mixers are also available for use with foams produced from two powders, "A" and "B."

(National Foam System, Inc., West Chester, Pa.)

Figure 8-4. Foam-making proportioner for a line of hose to use foam solution from a can. Line proportioner makes foam discharge by mixing air with the solution.

provision has been made for the control of flammable liquid fires, the private fire brigade members should be able to cope with these fires as readily as with any other type of fire. A fire in bulk storage of gasoline in tanks is often an easier fire to control than a fire in a lumber yard.

It is desirable that members of the fire brigade become familiar with every location in their plant in which flammable liquids are stored or used. These include locations of dip tanks, quenching tanks, gasoline and oil storage containers and tanks. Many flammable liquid hazards can best be controlled by automatic or fixed protection, including automatic sprinklers, foam extinguishing systems, carbon dioxide and dry chemical extinguishing systems, halogenated extinguishing agent systems, water spray equipment, automatic covers for dipping and quenching tanks, and drain devices to draw off flammable liquids.

Plant fire brigade members should work with and not against the automatic fire protection equipment provided. Automatic sprinklers should not be shut off indiscriminately when a flammable liquid fire occurs. In general, it should be assumed that the persons installing the sprinkler system or the processes employing the flammable liquids took into account the possible action of the sprinkler equipment. Care should be taken not to overflow containers of burning flammable liquids with a stream of water or to direct a solid stream into a flammable liquids container so as to spatter the liquid and spread the fire. In many cases such carelessness has caused burning flammable liquid to overflow and float on top of the water over a considerable area, frequently overtaxing the sprinkler system or other fire protection facilities. However, water in spray or fog form from hose streams is the best general fire fighting medium for oils and many flammable liquids.

Fires involving storage tanks of gasoline and other flammable liquids may present some hazard to fire fighters unless the tanks are equipped with adequate pressure relief valves. Internal pressures will be generated if a fire is burning all around a tank. However, if the tank has proper vents, the gases formed will blow off and burn. If the vents are too small and the pressure continues to increase, the tank may eventually rupture with explosive violence. The weakest part of a cylindrical tank usually is the end. In the case of large vertical tanks, the roof may blow off or the roof seams may open, releasing a blast of flame. This will ordinarily be well above the heads of the fire fighters. Usually the danger in such cases is from radiated heat. In the case of horizontal tanks, the ends may blow out. The entire tank may even be shot endwise for some distance. When working on a fire around a horizontal tank, it is best to stay at the side rather than in line with the end of the tank. Even where vents are not adequate, the building up of excessive pressure can often be prevented by cooling the tank with hose streams.

Foam Equipment

The manufacture of foam fire extinguishers was discontinued in 1969, but many units may still be in use today. Foam extinguishing systems are available for manual or automatic operation. Relatively simple systems are used for the protection of dip tanks and for the protection of small isolated storage tanks in an industrial plant. More elaborate systems are used in the protection of oil refineries, oil storage farms, fuel storage and handling installations, at airports, and at chemical and other manufacturing plants.

Figure 8-5. Roof-mounted foam generator.

In addition to the fixed foam systems installed to cover special hazards, fire fighters are frequently provided with portable foam generators designed for use with regular fire department hose equipment. Some of the portable foam equipment is in the nature of a "generator" or mixing hopper to be connected into the line of hose. Cans of foam-producing ingredients are dumped into the hopper, from which they are drawn into the hose where the foam is produced. Some generators are arranged for use with single foam-producing agents, and others have a double hopper providing for use of two foam-producing agents, which are combined when drawn into the hose.

Another type of portable equipment used for producing foam streams includes "proportioners" or specially-designed foam-producing nozzles having a simple syphon device using a pick-up tube, which, when inserted in a can of foam liquid, draws the correct proportion of the liquid from the can needed to form a foam stream.

Portable foam stream equipment that has to be set up hurriedly in an emergency may not be expected to be as efficient as fixed foam systems engineered to protect a particular hazard. Nevertheless, there are many situations when mobile foam streams fill an important need — for example, when burning flammable liquids have escaped from their containers. Frequently foam is spread over spilled flammable liquid to minimize the danger of ignition.

In operating foam streams, the foam must be flowed on the surface of the burning liquid. If the stream is played into the fire a foam blanket is not so readily formed.

So-called "fog-foam" nozzles have been developed to project the foam in a spray pattern, which, while having less range than a solid foam stream, may permit a more gentle application. Running or flowing flammable liquid usually has to be contained by a dike so that a foam blanket may be applied.

Foam should be flowed on a burning surface with reasonable rapidity, because foam applied to a large fire at too slow a rate will be broken down by heat and and its effectiveness greatly diminished. Rate of foam application is often more important than total quantity. If there is only a limited amount of foam powder available, it may be better to use it in several generators operating simultaneously rather than operating a single generator for a long time.

Fires in certain flammable liquids, such as alcohols, esters, and ketones, may not be controlled by some types of foam. Special alcohol-compatible foams have been developed for use on these fires.

Equipment is also available for developing and applying high expansion foam. There are three types of systems available — total flooding, local application, and portable systems. Total flooding systems can be used where there is an enclosure surrounding the hazard that will contain the foam. Local application sys-

(Gamewell Div., E. W. Bliss Co., Newton, Mass.)

Figure 8-6. Nozzle for high expansion foam.

(National Foam Systems, Inc., West Chester, Pa.)

Figure 8-7. Portable unit for high expansion foam.

tems are used to protect specific hazards where total flooding is unnecessary or impractical. There are two types of foam generators used in these systems — an aspirator and a blower. Foam expansion ratios range from 100:1 to approximately 1,000:1. It is important to use only those foam concentrates recommended for a specific system.

Carbon Dioxide, Dry Chemical, Inert Gas Equipment

In addition to carbon dioxide and dry chemical extinguishers, there are systems using the same general equipment for fire extinguishment on a larger scale and inert gas systems that may be used to prevent ignition of flammable vapors and gases.

Carbon dioxide storage systems are of two general types. The gas may be stored in high pressure cylinders or stored in refrigerated vessels at low pressures. The low pressure containers usually have warning devices to indicate if temperatures are getting too high. Low pressure carbon dioxide units have been installed in fixed installations for industrial fire protection.

Both carbon dioxide and dry chemical systems provide a snuffing effect with a wallop designed for the hazard protected. In using such equipment, do not assume that a fire is under control the moment the flame disappears. Application of the gas should be continued long enough for some cooling effect to develop. In many cases, the flames are extinguished almost instantly, but sufficient heat remains to cause a dangerous reflash. Also, it is well to remember that the snuffing action excludes oxygen from the fire area. In an area filled with the extinguishing gas, a fire fighter could be suffocated by lack of oxygen. This would not be likely from the amount of gas available from hand appliances, but some of the fixed systems are designed to completely flood a fire area with gas.

Halogenated Agent Systems

Another type of extinguishing system that is effective against flammable liquid fires uses a halogenated extinguishing agent — Halon 1301 or Halon 1211. Such systems are also useful against flammable gases and where live electrical equipment exists. Halon systems are available for total flooding, local applications, and special applications. Although relatively low concentrations of halons are needed for extinguishment, there is a toxicity hazard to personnel in the area where they are used. The hazard involves the products of combustion, the halons themselves and the halon decomposition products evolved when the agent comes in contact with the heat of a fire. Low concentrations

(Walter Kidde & Belleville, N.J.)

Figure 8-8. Portable unit for producing high expansion foam and foam discharge.

Figure 8-9. Carbon dioxide storage cylinders in a high pressure system.

(Cardox Div., Chemetron Corp.)

Figure 8-11. Carbon dioxide extinguishing system for a foundry operation with both manual and automatic operational control.

of the agents can be tolerated for short periods of time, but the more toxic decomposition products intensify the hazard. Do not linger in an area where halon has been discharged, especially if the discharge was from a total flooding system.

Water Spray Systems

A fine water spray has a cooling and blanketing effect on flammable liquid fires very different from the effect of water applied in a solid stream. Water spray extinguishing equipment, consisting of fixed spray nozzles available with various types of discharge patterns and connected to a suitable water supply by permanent piping, is effective in controlling many flammable liquid fires. The spray nozzles may be either the open or the automatic type. The advantage of the permanent system is that the nozzles are arranged to cover the desired area at one time. Even automatic sprinkler systems operating under adequate pressure provide a desired blanketing effect useful in controlling many flammable liquid fires and in dissipating heat.

GAS FIRES

Most kinds of fires should be extinguished as promptly as possible. However, such is not the general rule for gas fires. They should be allowed to burn until the gas flow can be cut off. Meanwhile, the surroundings should be sprayed with water to absorb heat and prevent ignition of combustible materials in the vicinity. It is proper practice to extinguish a gas flame momentarily if this is necessary to permit reaching and operating a shutoff valve.

Figure 8-10. Insulated storage vessel for low pressure carbon dioxide system.

Figure 8-12. Water spray equipment installed to protect a chemical process system.

(Grinnell Co., Providence, R.I.)

Figure 8-13. Water spray protecting transformers.

Gas flames can be extinguished by snuffing with carbon dioxide, a halogenated extinguishing agent, or in other ways. However, if the unburned gas continues to flow, it is likely to form an explosive mixture with air and sooner or later reach some source of ignition. The explosion that follows will probably cause much more damage than would have the original gas fire had it been allowed to burn.

ELECTRICAL FIRES

When fighting fires in electrical equipment, consider the danger to yourself and possible damage to the equipment. Whenever possible, first shut off the current, and then ground the equipment. Sometimes a dangerous potential remains after the current has been shut off; grounding the equipment will discharge it. If the fire continues to burn, use the extinguishing means most suitable for the conditions.

For small fires involving wire and cable insulation, portable extinguishers charged with halon, carbon dioxide, or a dry chemical are satisfactory, and will not damage electrical equipment. Since these extinguishing agents are nonconductive, they may be used safely on live electrical equipment, provided you are not within arcing distance of the equipment.

If the fire has spread appreciably, it is better to apply water as quickly as possible by means of spray nozzles. Sometimes larger spray streams are required, but ordinarily a small stream is preferable, because it produces less water damage. If the water is relatively free of minerals and contaminants, it may be possible to dry the equipment out with little or no damage beyond that caused by the electric current and fire. If the electricity is not shut off before extinguishment operations are begun, additional damage may result.

Do not use soda acid, calcium chloride, foam, or loaded stream extinguishers on fires in electrical equipment unless no suitable agent is available. These liquids leave solid and corrosive deposits, which will cause additional damage. If these agents must be used, always cut off the current and ground the equipment before attempting to extinguish the fire; otherwise these electrically conductive liquids will become a shock hazard.

Oil switches, oil-filled transformers, and other electrical equipment containing oil involve the additional hazard of oil fires. The oil used has a relatively high flash point, but it may be heated and ignited by excessive current or by an electric arc. After the current has been cut off, such fires can be extinguished by any of the several methods of extinguishing oil fires, or by water fog or spray streams. Many such switches and transformers are equipped with permanently installed water spray nozzles.

Preventing Electrical Accidents

A word of caution is needed regarding electrical fires and accidents. There is more danger of fire fighters becoming seriously shocked by stepping into water charged by fallen wires than by playing a stream inadvertently against an overhead wire. Too much dependence for insulation against electrical accidents should not be placed in the wearing of rubber boots. Boots are often made of rubber containing considerable carbon and other substances that may permit the passage of electrical currents.

Similarly, serious accidents have resulted from too much confidence in, or improper use of, rubber gloves, which may have deteriorated due to improper handling and storage. Even a tiny pinhole may be enough to permit a fatal charge to pass through to the wearer. Gloves should be carefully stored and protected against

(Ansul Co., Marinette, Wis.)

Figure 8-14. Combined agent application technique. Twin nozzle discharges dry chemical (foreground) and aqueous film forming foam from other nozzle secures a flammable liquid area.

(GMC Truck & Coach Div., Pontiac, Mich.)

Figure 8-15. Intraplant fire truck carrying a selection of special extinguishing equipment.

excessive heat. Fire fighters should leave the use of rubber gloves to the trained crews of the electric utility.

Removing a Victim from a Live Wire

Many would-be rescuers have paid with their lives for attempts to pull victims away from live electrical wires or equipment barehanded. Whenever possible, it is desirable to shut down the current before attempting rescue operations. However, if this requires more than a few moments, some other form of rescue attempt will be necessary in order to prevent fatal injuries and permit immediate efforts to restore breathing before suffocation results. Obviously, if a victim is in contact with a source of electricity having a high potential, any rescue operations are likely to be dangerous because of the danger of the rescuer forming an electrical path to ground, even without coming into actual physical contact with the victim or with the energized equipment responsible for the accident.

In removing a victim from a live wire the following steps are usually recommended.

(*a*) Wear rubber gloves or some other gloves if available. Use rubber boots for any additional insulation they may provide.

(*b*) Stand on dry wood or other insulating material, such as a rubber salvage cover, to prevent an easy ground connection. A short wooden ladder will provide a good platform to stand on.

(*c*) Throw a crowbar or some other metal object across the wire on the side between the victim and the source of the current in order to ground as much of the current as possible. Where available, a fire department plaster hook or pike pole can be used to drive or hold the wire to the ground. The dry wooden handle provides insulation for the person placing the pike pole. Observe all other precautions mentioned, such as standing upon some nonconducting substance and wearing rubber boots and gloves if available.

(*d*) Do not touch the victim with bare hands as long as there is a possibility that the current is still on. It is advised that he or she be pulled or rolled off the wire by means of a long dry wooden pole.

Cutting Wires

Instructions given to most fire departments regarding situations in which outside electrical service wires must be cut are to leave this work to the electrical utility emergency crews available and equipped for this work. Usually only power company linemen are able to judge the carrying capacity of an unidentified electrical conductor by observing the type of conductor used. Industrial plants usually have a chief electrician or other person familiar with the electrical service within the plant. The fire brigade should have this person explain to its members the possible places where electrical accidents might occur in the plant and how these accidents may be guarded against under fire conditions.

Rarely should it be necessary to cut electrical wires at fires. Usually, shutting off individual circuits or services will achieve the desired results. Even then, care must be taken not to interrupt essential electrical equipment unless absolutely necessary. The only cases where it is necessary to cut outside electrical service lines to a building would be where the building is seriously involved in fire. In such cases, the utility crew, if called promptly, usually will be on hand in time to handle the situation. In no instances should any utility wires be cut without approval or instruction from the utility concerned.

When it becomes necessary to cut electrical services, the wires are severed close to a pole or support on the incoming side of the circuit in order that live wires will not dangle to the ground. Only persons skilled in wire cutting and having the proper tools should attempt this work. Care must be taken to have all other persons remain at a safe distance when wires are cut, as there may be a dangerous whipping effect as some wires are released. Fire fighters must take care not to needlessly interrupt electrical service that is needed for life safety or to operate fire protection equipment such as fire pumps.

METAL FIRES

Fires involving certain combustible metals present different problems. Most of the conventional extinguishing agents are ineffective against burning metals. Some may intensify burning or introduce an explosion hazard. Factors influencing the choice of extinguishing agent include the type, quantity, and form (shavings, fabricated parts, or molten metal) of the burning metal.

For example, a small amount of water applied to burning magnesium chips will accelerate combustion. However, a large amount of water applied to a small number of chips or to fabricated parts can cool the metal and extinguish the fire. Water should not be used on molten magnesium, because steam may form and escape with explosive force, scattering the burning molten metal throughout the area.

Fire brigade members should know what combustible metals are used in the plant and what extinguishing agents and techniques to use on them. It is important that the brigade practice extinguishing fires in these materials in order to become familiar with their fire behavior.

Suggested Reading

Fire Protection Handbook, 14th Edition, 1976, NFPA, Boston.

NFPA 12, *Standard on Carbon Dioxide Extinguishing Systems*, 1973, NFPA, Boston.

NFPA 12A, *Standard on Halogenated Fire Extinguishing Agent Systems — Halon 1301*, 1973, NFPA, Boston.

NFPA 12B, *Standard on Halogenated Fire Extinguishing Agent Systems — Halon 1211*, 1973, NFPA, Boston.

NFPA 14, *Standard for the Installation of Standpipe and Hose Systems*, 1976, NFPA, Boston.

NFPA 17, *Standard for Dry Chemical Extinguishing Systems*, 1975, NFPA, Boston.

Fire Hose Practices, 6th Edition, 1974, International Fire Service Training Association, Stillwater, Oklahoma.

Chapter 9

PLANT FIRE PROTECTION SYSTEMS

Previous chapters have described fire extinguishers and hose. These are items of fire protection equipment provided in most properties. This chapter is devoted to identifying protection that is specifically provided for the one property in which fire brigade members must operate.

WATER SYSTEMS FOR FIRE PROTECTION

Fire fighters must know about the water system that protects the property. They should know where the water comes from and how it is to be used. Fire fighters should make at least rough sketches of the plant's water system for fire protection, embodying the details applying to the property they are assigned to protect.

Wherever there is a public water system, that system is often the principal source from which a factory or other property obtains water for its fire protection. In the simplest case, the private property has no water system of its own, and the water for fire fighting must be taken from hydrants on the public system. In a typical arrangement, water comes from a large water main in the city system through a branch main down the side street to bring water to the site of the property.

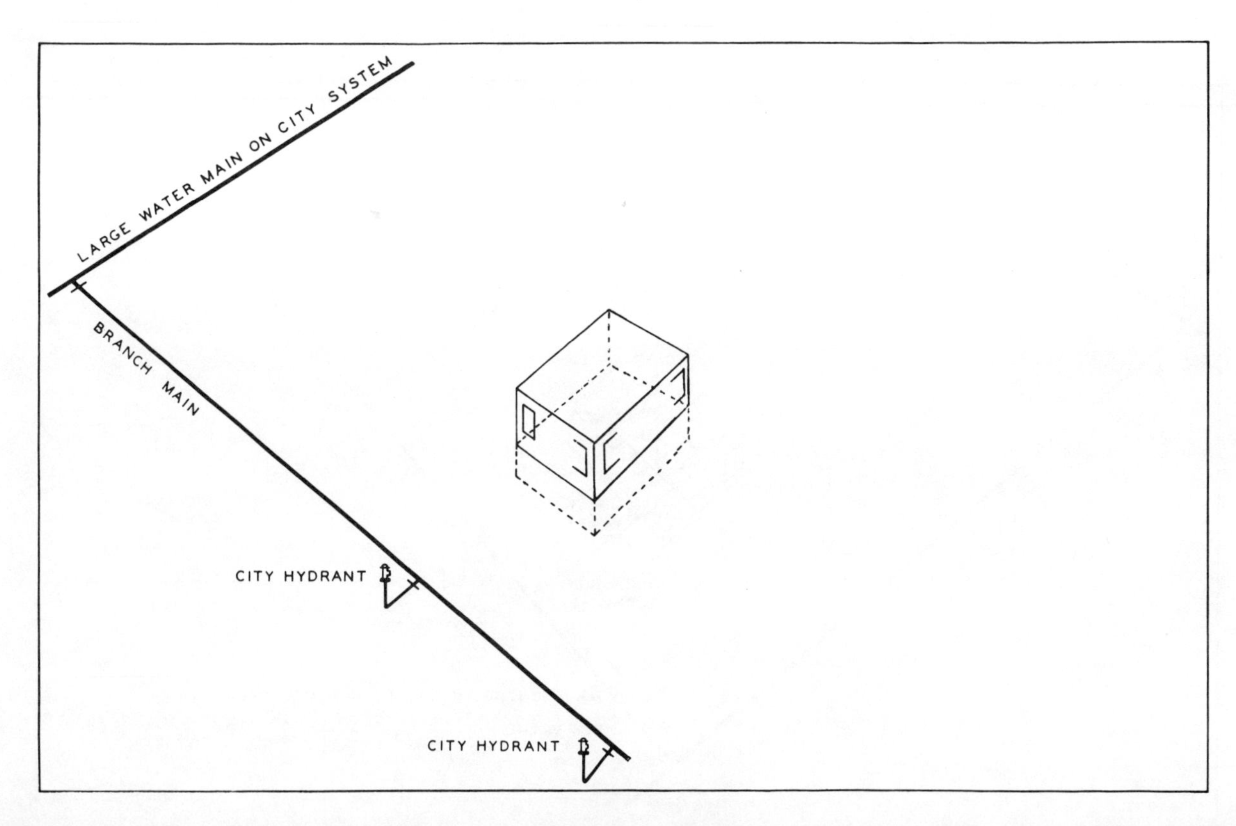

Figure 9-1. In a simple case, the private property has no water system of its own, and water for fire fighting must be taken from hydrants on the public system.

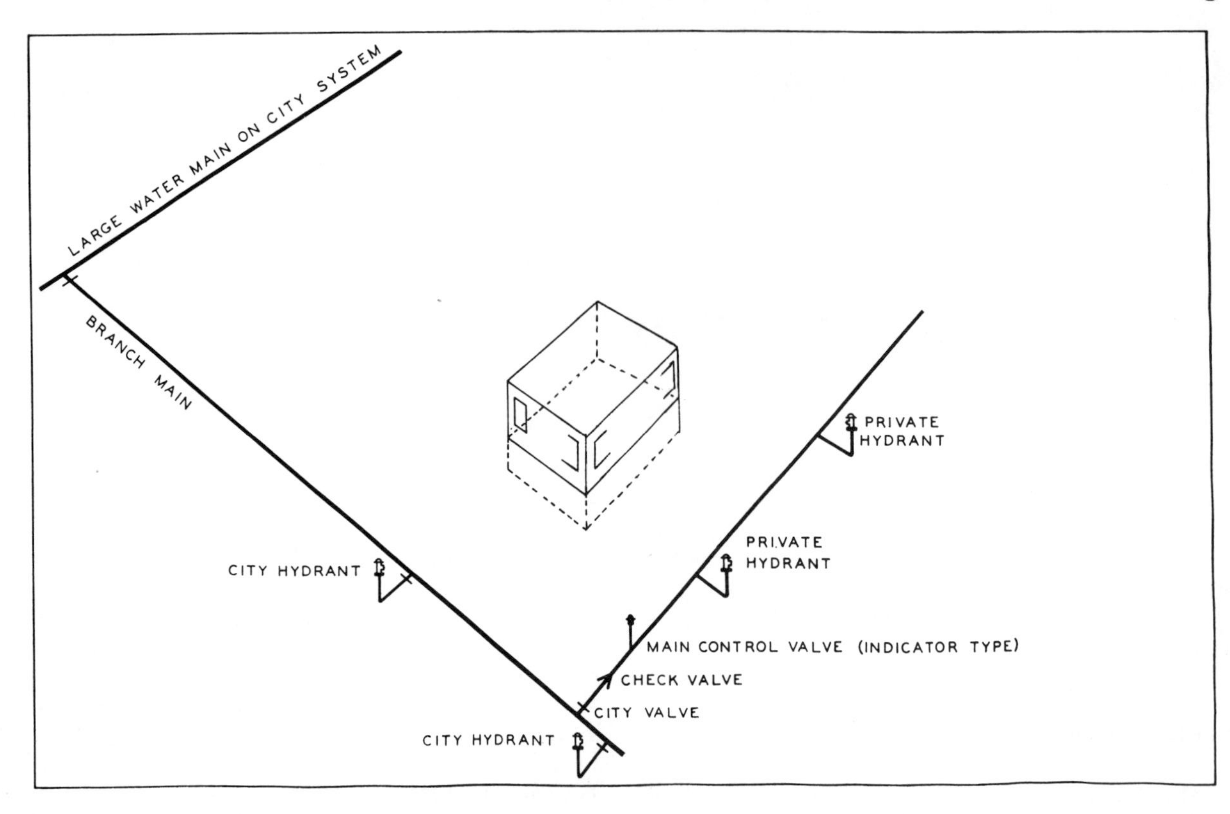

Figure 9-2. Where the property has piping of its own for fire protection, private piping is connected to the branch main on the public system.

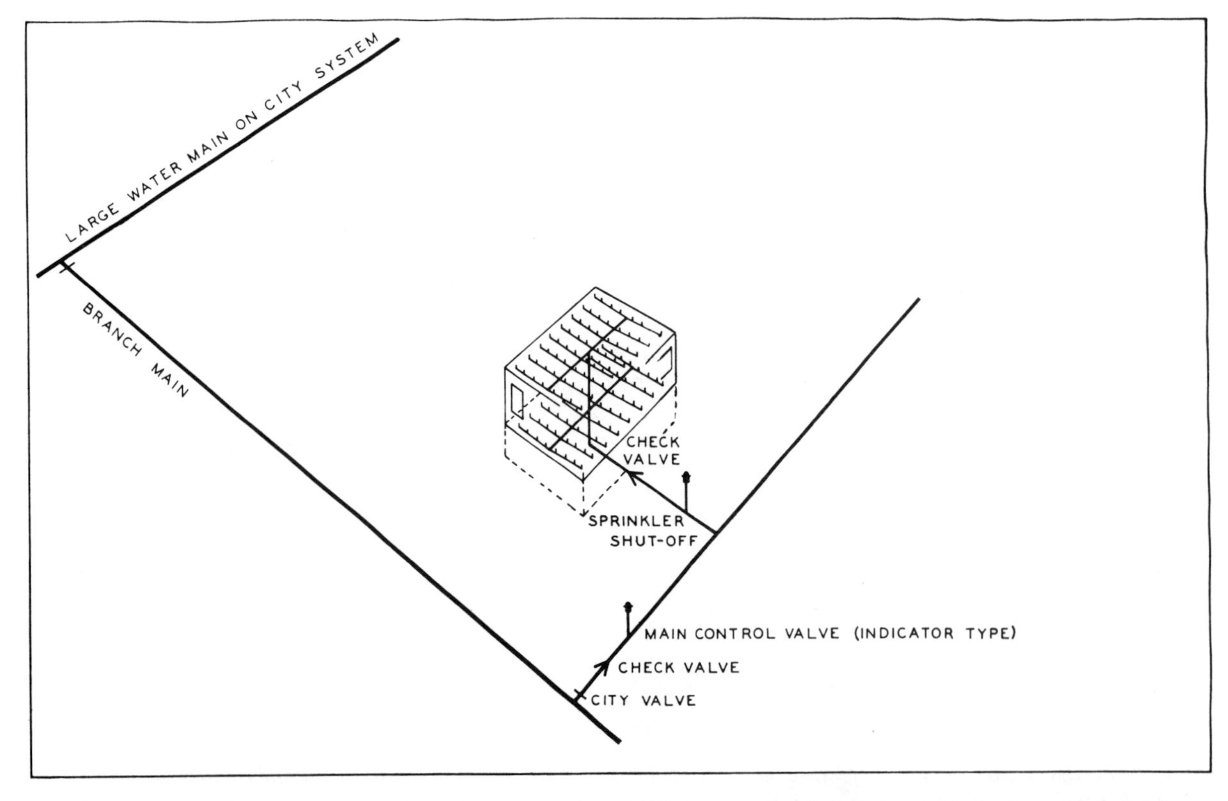

Figure 9-3. The principal components of an automatic sprinkler system supplied with water from the public system.

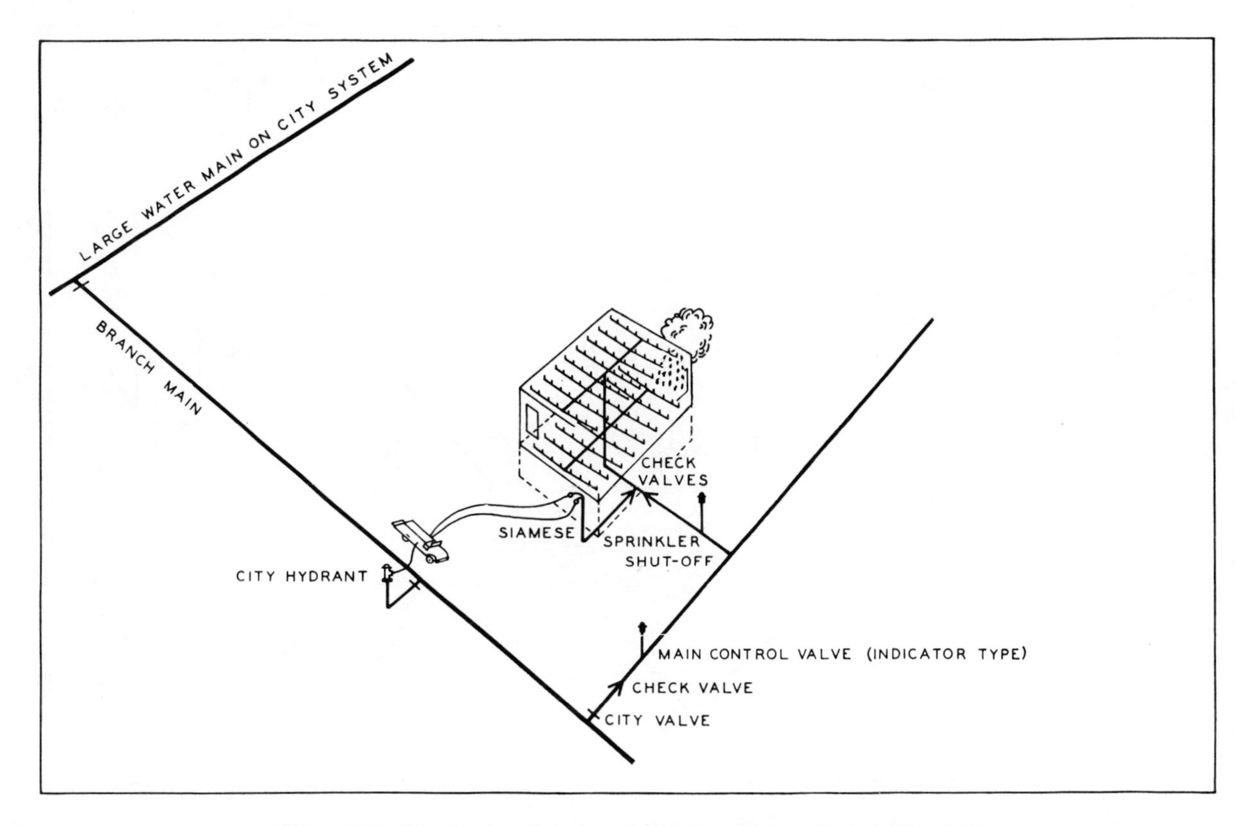

Figure 9-4. The fire department connection to an automatic sprinkler system.

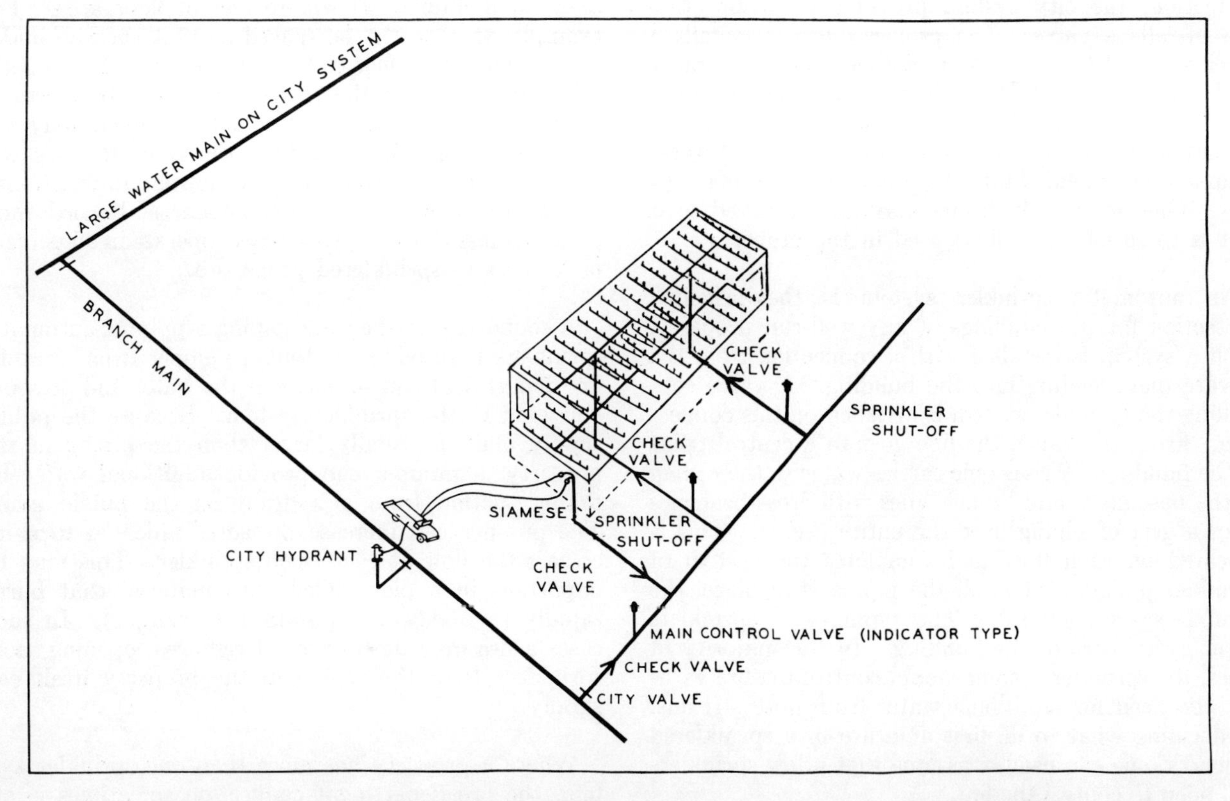

Figure 9-5. Arrangement of fire department connection where there are two or more sprinkler risers.

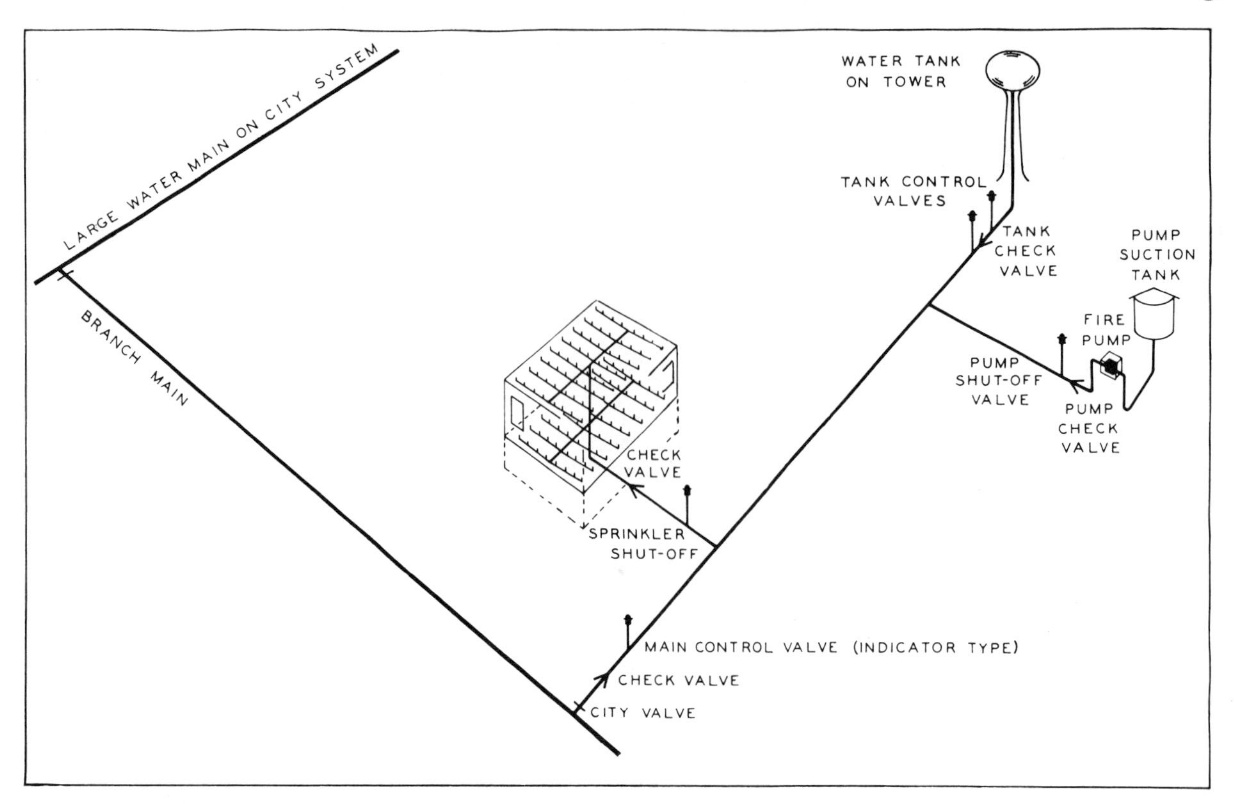

Figure 9-6. Water supplies provided in a property — tanks and fire pump.

Where the property has piping of its own for fire protection, the city system provides the connection and installs a valve. The property owner installs a check valve at this point and another valve (preferably of the indicator type) that shows at a glance whether it is open or closed. This is the main control valve of the supply from the public system. The private water main is then extended into the property as far as necessary. One or more hydrants may be connected to it if it is to supply hose lines used in the property.

An automatic sprinkler system is the principal protection for the buildings of any well-run property. Such a system is installed with a connection from the private main leading into the building. A valve controlling the sprinkler system is installed on this connection. From the valve, the line goes to a central point in the building. A riser pipe carries water to the ceiling of the basement, and branch lines with cross branches form a grid of piping over the entire ceiling. This is repeated on each floor and completes the system of sprinkler piping. When all the piping is in place, the sprinklers are installed. The piping and sprinklers cover every part of the building. In the majority of cases, the sprinkler system should control the fire without the need for additional water from hose streams. In deciding what to do first at a fire in a sprinklered property, one can usually assume that a few sprinklers will open to control the fire.

Sprinklers are spaced so that each will cover a designated number of square feet of floor area. For example, they might be spaced so that each sprinkler covers 100 square feet (9.3 square meters). It is a common misconception that, when one sprinkler operates (fuses), they all operate. In fact, however, only the sprinkler or sprinklers in the vicinity of the fire will operate and apply water in a somewhat umbrella-like pattern to cover the designated floor area. Records indicate that less than three sprinklers operate in 95 percent of all fires in sprinklered properties.

A connection to the water piping supplying automatic sprinklers is provided so that a pumper from a public fire department can supplement the water and pressure delivered to the sprinkler system. Because the public supply main is usually larger than the piping in the property, a pumper can provide additional water by taking suction from a hydrant on the public main. The pumper can increase pressure, which in turn increases the flow at each open sprinkler. This may be important in a plant filled with material that burns rapidly (a woodworking shop, for example). In such cases a fire may flash over a large area, opening more sprinklers than the piping in the property itself can supply.

Where a property has more than one sprinkler system, the fire department connection sometimes is ar-

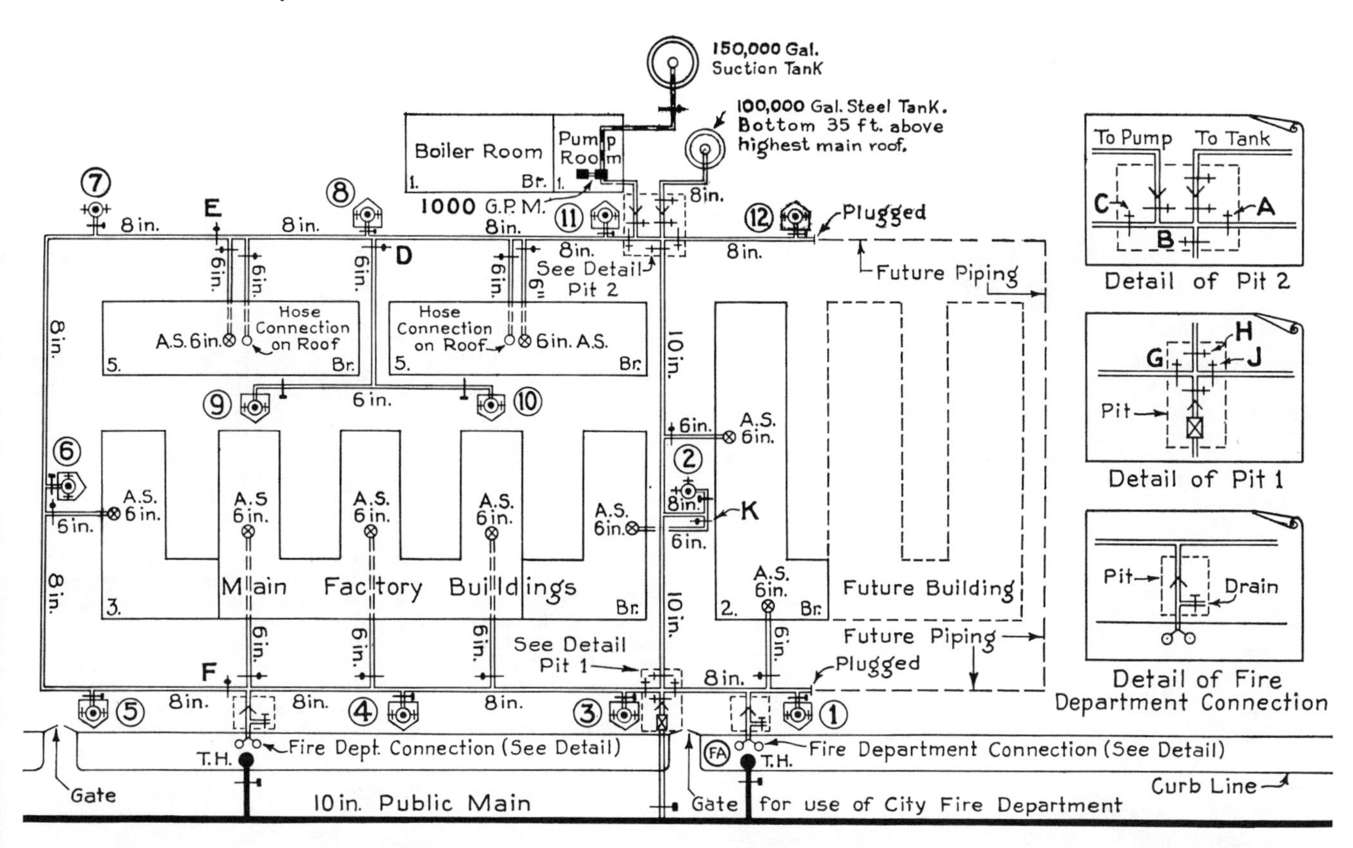

Figure 9-7. A plan like this is available for most plants. It shows looped yard mains, water supplies, valves, and hydrants.

ranged so that it can supply risers on two or more sprinkler systems.

The water system for fire protection installed within the property itself is most reliable when it has its own supply of water in case the supply from the public system is interrupted. The basic method of providing water at a property is to use a storage tank with a permanently installed fire pump. The storage tank (or reservoir) can be made to store water for the expected fire flow for any number of hours. Gravity tanks are also used as a method of providing a reliable supply at adequate pressures.

The accompanying diagrams show details of protection such as are found in the usual industrial plant. Since you are particularly concerned with your own plant, see how far this general information applies to it. A plan of the buildings and water supply system of the plant made by insurance engineers usually is available and is useful for the purposes of this study. On it may be traced the sources of water supply (street connections, tank, pump) and the arrangement of yard mains, valves, and hydrants. By studying a plan of the plant in which you work, you can best see what protective equipment you need to know about.

The principal parts of the water system for fire protection should be identified.

(*a*) Valves in the water system should generally be of an indicating type so that one can readily determine if the valve is open or closed. One type of indicating valve is the outside screw and yoke gate valve. Gate valves are provided with indicator posts for use in underground piping. Another type of indicating valve is the butterfly valve. A check valve is used where water is intended to move in one direction only. A pumper can boost pressure because there is a check valve at the connection between the street main and the yard main.

(*b*) Private hydrants may differ from those used on the public water system. The pressure on private hydrants may come from a gravity tank or private pump. An automotive pumper is not needed to supply hydrant lines for, where extra pressure is required, the private fire pump can be started. For these reasons, private hydrants usually have no large suction outlets. On water systems laid out as illustrated, it would be bad practice to try to take suction with a public fire department pumper from a private or yard hydrant. The best practice is to equip private hydrants with valves on each outlet, for in this way each hose line can be turned on or shut off individually.

(*c*) Elevated tanks provide pressure due to gravity. Pressure tanks are sometimes used in factories but less commonly than gravity tanks.

(*d*) Permanently installed fire pumps may be driven

(Mueller Co., Decatur, Ill.)

Figure 9-8. Outside screw and yoke gate valve. The illustration shows valve in closed position. If the valve were open, the stem would stick up above the wheel, so its position can easily be determined. When used in an underground system, such valves are installed in pits.

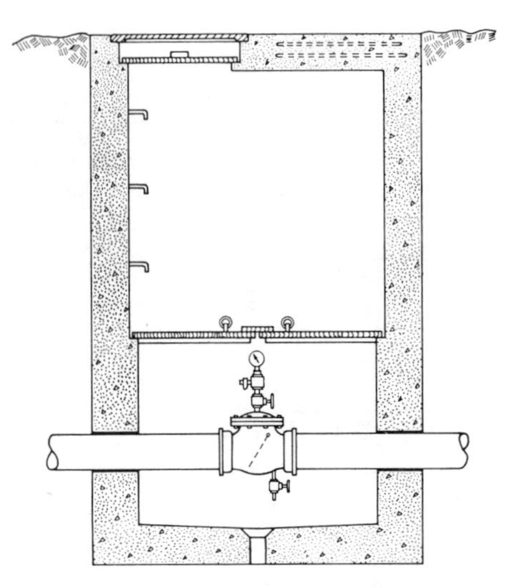

Figure 9-10. Customary installation of a check valve in a valve pit. Outside screw and yoke gate valves can be similarly installed. Reinforced concrete pit shown is most substantial design for such pits and, in the long run, the most economical. It is the preferable design when a pit must be made tight against ground water. Pits can also be constructed with concrete blocks or precast concrete sections with gravel in the bottom where soil conditions permit constant drainage. Pits can be framed in wood for temporary locations or where treated lumber is used. The pit must have a reinforced top. A manhole cover can be a standard assembly, but it must be provided with a double cover for protection against freezing. In severe freezing conditions, the lower cover illustrated may need to be used.

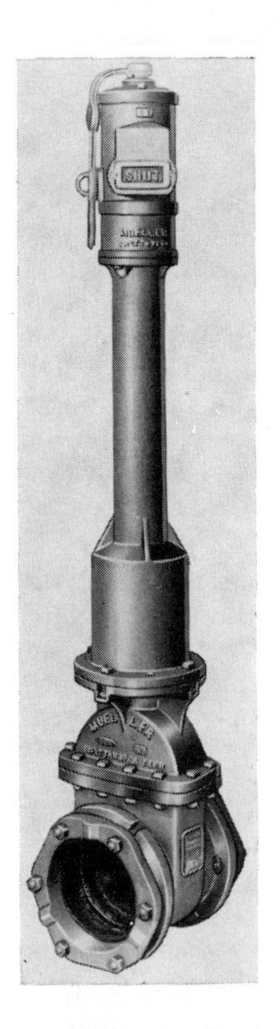

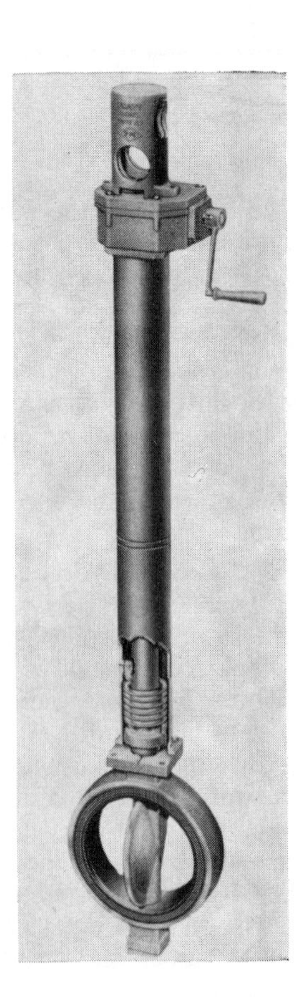

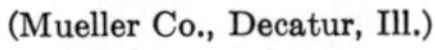

(Mueller Co., Decatur, Ill.) (Henry Pratt Co., Aurora, Ill.)

Figure 9-9. Valves for use in underground systems. (Left) Gate valve with indicator post and (right) butterfly valve with indicator. In both cases, a target shows if valve is open or closed.

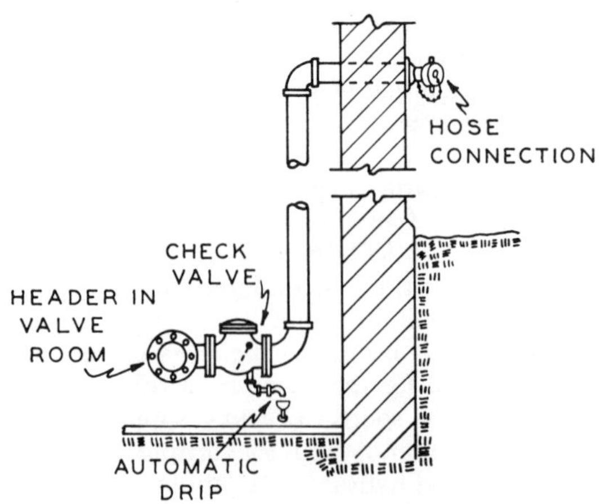

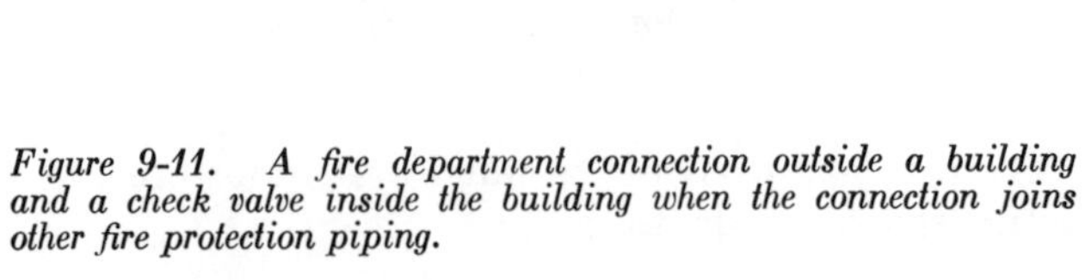

Figure 9-11. A fire department connection outside a building and a check valve inside the building when the connection joins other fire protection piping.

(Chicago Bridge and Iron Co., Chicago, Ill.)

Figure 9-12. Elevated tanks supplying water for private fire protection. (Top) Wood stave tank on a factory roof, and (bottom) a steel tank on a 100-foot column.

(Fairbanks, Morse & Co., Kansas City, Kansas)

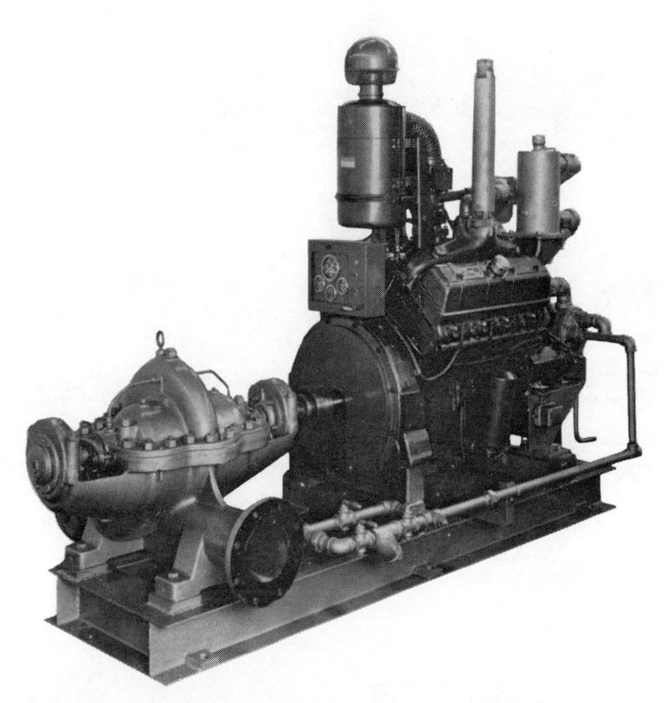

(FMC Corp., Peerless Pump, Indianapolis, Ind.)

Figure 9-13. Fire pumps for permanent installation in buildings — electric motor drive (top) and internal combustion engine drive (bottom). Pumps are also made for vertical shaft drive.

by electric, steam, or internal combustion engine power. These pumps take suction from reservoirs or tanks.

(*e*) Automatic sprinkler systems are the most important single item of private fire protection. Records of over 80,000 fires show that 96 percent were controlled by sprinklers. In most of the few failures, the valve was found closed. The records also show that in over 73 percent of all sprinklered fires fewer than five sprinklers opened.

(*f*) Standpipe and hose systems carry water to hose connections on each floor of a building by vertical pipes.

Figure 9-14. Automatic sprinkler fusible linkage (left) at the start of operation (right).

(Grinnell Co., Inc., Providence, R.I.)

Figure 9-15. Sprinkler installation in an industrial warehouse.

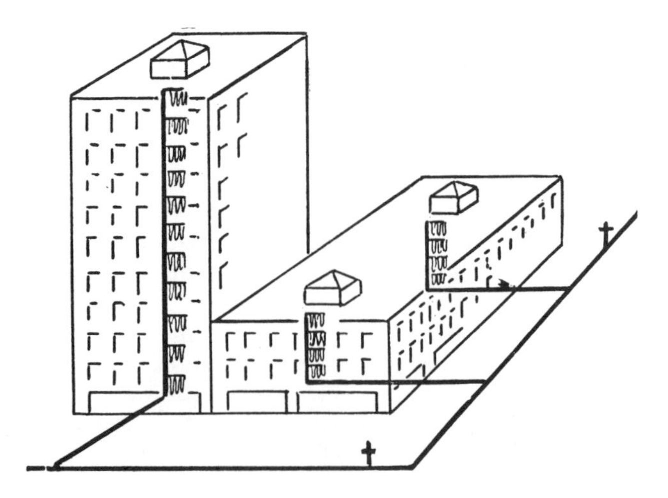

Figure 9-16. Standpipe and hose systems. Standpipes, usually supplied from street mains, furnish water for fire fighting in buildings. Hose sets are provided at outlets on each floor.

SPECIAL HAZARD PROTECTION

Many plants have hazards that require special protection. Industrial fire fighters should familiarize themselves with such protection in their own plants. Fog or spray, foam, carbon dioxide, halons, or dry powders are used as extinguishing agents in this equipment.

FIRE ALARMS

Most plants have an arrangement for sounding a fire alarm to notify brigade members in various parts of the property. In some cases, the alarm may also call brigade members from their homes during hours when the plant is closed. A common practice is to divide the plant into fire areas and assign each a coded signal. Where this practice is followed, brigade members are given cards showing the coding and location of each fire area. The method of sounding the alarm may be by horn, whistle, or voice communication.

When a plant is located in a built-up part of a city, there is usually a municipal fire alarm box located at the plant or on the street nearby.

(Bliss-Gamewell Div., Natick, Mass.)

Figure 9-17. Street fire alarm box on public fire alarm system located at a plant.

(S. H. Couch Co., North Quincy, Mass.)

Figure 9-18. Manual fire alarm box for use in private property.

There are two other fire alarm services that can be used in places where, in the opinion of the superintendent of the public fire alarm system, such service is appropriate. One possible arrangement involves a number of auxiliary alarm boxes located throughout the property and connected to a master box, which is tied into the municipal fire alarm system. If any one of the auxiliary boxes is tripped, the master box signal will be automatically transmitted to the fire department. Another possibility is a connection permitting direct transmission of the sprinkler system water flow alarm to the public fire department.

Large properties frequently have a central alarm facility of their own. The plant central system provides a point not only for receiving alarms, but also for supervising plant guards and patrolmen.

Few plants, however, are large enough to warrant operating their own central stations, and the large majority use the services of private central stations, which are available in principal cities. These central stations provide supervision for a wide range of details in a large number of properties and give very reliable

(S. H. Couch Co., North Quincy, Mass.)

Figure 9-19. Annunciator panels are alarm equipment for indicating at a central point in the plant the existence of various conditions: for example, operation of manual fire alarm boxes or thermostats, waterflow in sprinkler lines, abnormal pressures, temperatures of voltages in equipment.

(Autocall Co., Shelby, Ohio)

Figure 9-20. Annunciator and recording equipment for waterflow alarms, indication of valve operation, or other signals, can be designed for any service required. For example, a control cabinet can be provided for a great variety of circuits from sending devices anywhere on the property. The circuit can store relatively complicated instructions for recording the signal and taking action on it. The stored circuits illustrated provide instructions for printing out a description of signal, date, time, and other coded information.

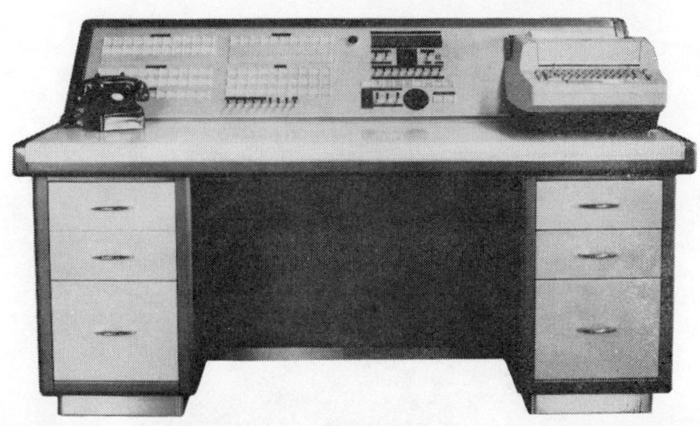

(Autocall Co., Shelby, Ohio)

Figure 9-21. A large plant may have an alarm control center to monitor a great variety of signals. The annunciator features are often incorporated in a control desk or console. The console illustrated provides automatic typewritten records of signals received and action taken by the operator in response to signals.

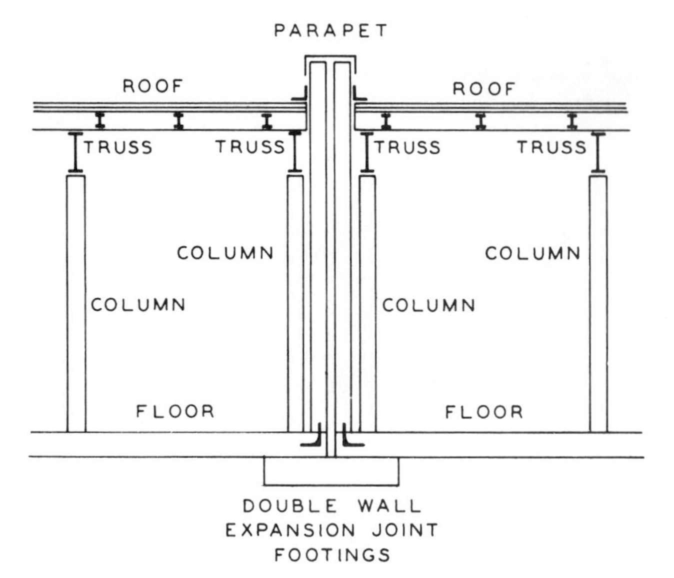

Figure 9-22. A fire wall of a type suitable to separate areas of a one-story industrial building.

Figure 9-23. Fire doors provide fire cutoffs where there has to be an opening in a fire wall. Doors shown are sliding, tin-clad type.

Figure 9-24. Window protection — wired glass in metal frames and open sprinklers.

service. Industrial plants find it an advantage to use central station supervisory service, because such central stations are operated by persons who can give their undivided attention to matters of plant security.

MEASURES TO RESTRICT SPREAD OF FIRE

Specific measures are taken in every building to restrict the spread of fire. The location and purpose of measures such as fire walls, fire doors, window protection, shutters, wired glass in metal frames and open sprinklers, and similar protective arrangements must be understood.

Suggested Reading

NFPA 7, *Recommendations for Management Control of Fire Emergencies*, 1974, NFPA, Boston.

NFPA 13, *Standard for the Installation of Sprinkler Systems*, 1976, NFPA, Boston.

NFPA 13E, *Recommendations for Fire Department Operations in Properties Protected by Sprinkler and Standpipe Systems*, 1973, NFPA, Boston.

NFPA 14, *Standard for the Installation of Standpipe and Hose Systems*, 1976, NFPA, Boston.

NFPA 20, *Standard for the Installation of Centrifugal Fire Pumps*, 1976, NFPA, Boston.

NFPA 22, *Standard for Water Tanks for Private Fire Protection*, 1976, NFPA, Boston.

NFPA 24, *Standard for Outside Protection*, 1973, NFPA, Boston.

NFPA 71, *Standard for the Installation, Maintenance, and Use of Central Station Signaling Systems*, 1974, NFPA, Boston.

NFPA 72A, *Standard for the Installation, Maintenance, and Use of Local Protective Signaling Systems for Watchman, Fire Alarm, and Supervisory Service*, 1975, NFPA, Boston.

NFPA 72B, *Standard for the Installation, Maintenance, and Use of Auxiliary Protective Signaling Systems for Fire Alarm Service*, 1975, NFPA, Boston.

NFPA 72C, *Standard for the Installation, Maintenance, and Use of Remote Station Protective Signaling Systems*, 1975, NFPA, Boston.

NFPA 72D, *Standard for the Installation, Maintenance, and Use of Proprietary Protective Signaling Systems for Watchman, Fire Alarm, and Supervisory Service*, 1975, NFPA, Boston.

NFPA 73, *Standard for the Installation, Maintenance, and Use of Public Fire Service Communications*, 1975, NFPA, Boston.

NFPA 80, *Standard for Fire Doors and Windows*, 1975, NFPA, Boston.

Water Supplies for Fire Protection, 2nd Edition, 1971, International Fire Service Training Association, Stillwater, Oklahoma.

Bahme, Charles W., *Fire Officer's Guide to Extinguishing Systems*, 1973, NFPA, Boston.

Chapter 10

FIRE FIGHTING PRACTICE with AUTOMATIC SPRINKLER SYSTEMS

Where a plant is protected by a well-designed and carefully maintained system of automatic sprinklers, the fire brigade's work is simplified in a majority of fires. The private fire brigade and the public fire department should plan operations to make this equipment effective. A mistake poorly trained brigades sometimes make is to lay hose lines for fire fighting in such a way that the hose streams take water that could better be put on the fire through the sprinkler systems.

In any property protected by automatic sprinklers, a principal responsibility of the fire fighting force is to make certain that the sprinklers are operating properly and are adequately supplied. In all cases where sprinklers operate on a fire, they should not be shut down until the officer in charge of fire fighting operations has determined that the fire is under control. Premature closing of sprinkler supply valves has been a leading cause of major fire losses in industrial and other important properties.

An equally serious cause of large fire losses in properties protected by automatic sprinklers has been the excessive use of hose streams from the water system supplying the sprinklers. This deprives the sprinklers of the volume of water and pressure required for their effective operation. Sprinklers are designed to put out the fire or at least hold it in check. The job of the fire department or fire brigade is to complete extinguishment, perform ventilation and salvage, and restore protection.

Restoring protection includes replacing fused sprinklers with sprinklers of the correct type and temperature rating. Fire departments as well as plant brigades responding to fires in sprinklered buildings should be equipped with an assortment of sprinklers, including the upright, pendant, and flush-mounted types. They should also be equipped with wrenches for changing all types of sprinklers. If the sprinklers are on a dry-pipe system, the dry valve must be reset and air pressure restored. In some plants, fire brigade officers are trained to do this. In others, sprinkler contractors are called. In the latter case, if the temperature is above freezing, the water may be directed into the piping to supply the sprinklers pending the arrival of the service company representative. In any event, protection is not considered restored until sprinklers are again ready for service and alarm systems have been restored.

Automatic fire protection equipment is designed to control fire without human help. Ordinarily, it can be depended on to hold a fire in check at the point of origin and give prompt notification, so that proper measures can be taken to make certain that extinguishment is complete. Shown is a typical factory building supplied by an automatic sprinkler system. If a fire occurs in this building, the sprinkler system, without any help from the fire brigade or the fire department, should control the fire. In deciding what to do first at a sprinklered fire, one usually may assume that a few sprinklers will open and control the fire.

Rule 1. Someone should be sent to the control valve to

(*a*) Determine promptly that the sprinklers are operating properly;

(*b*) Open the valve if it has been closed;

(*c*) Shut the valve promptly when ordered to do so by the officer in command of the fire fighting operations; and

(*d*) Remain at the valve so that, in the event of rekindling or any detected extension of the fire, the valve can be reopened promptly. It is very important for the fire fighter to remain at the valve until the protection is completely restored.

Rule 2. The first or second public fire department pumper to arrive should hook up to the fire department connection and supply the sprinkler system as illustrated. The pumper is connected to a city hydrant. In the situation illustrated, the hydrant is on a larger main, assuring a better supply than is available from one

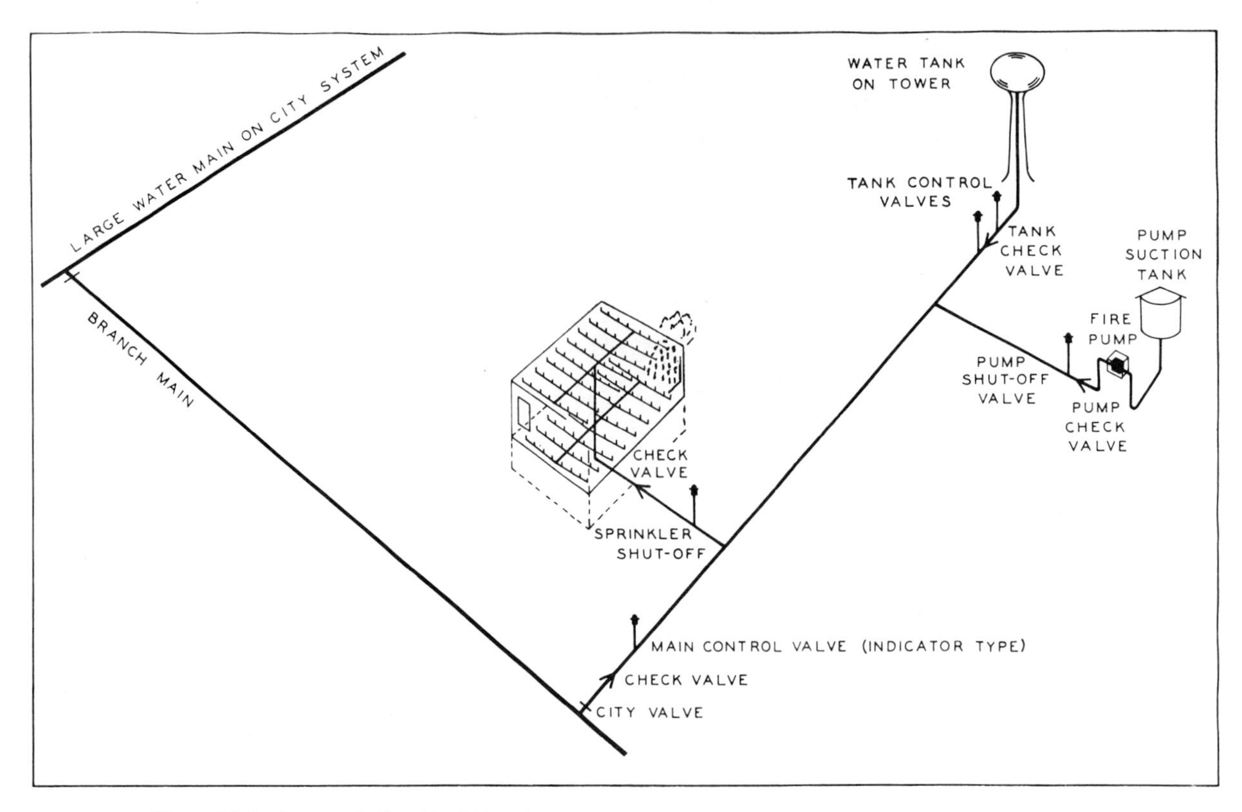

Figure 10-1. In a sprinklered building, it must be assumed that sprinklers will open and control the fire.

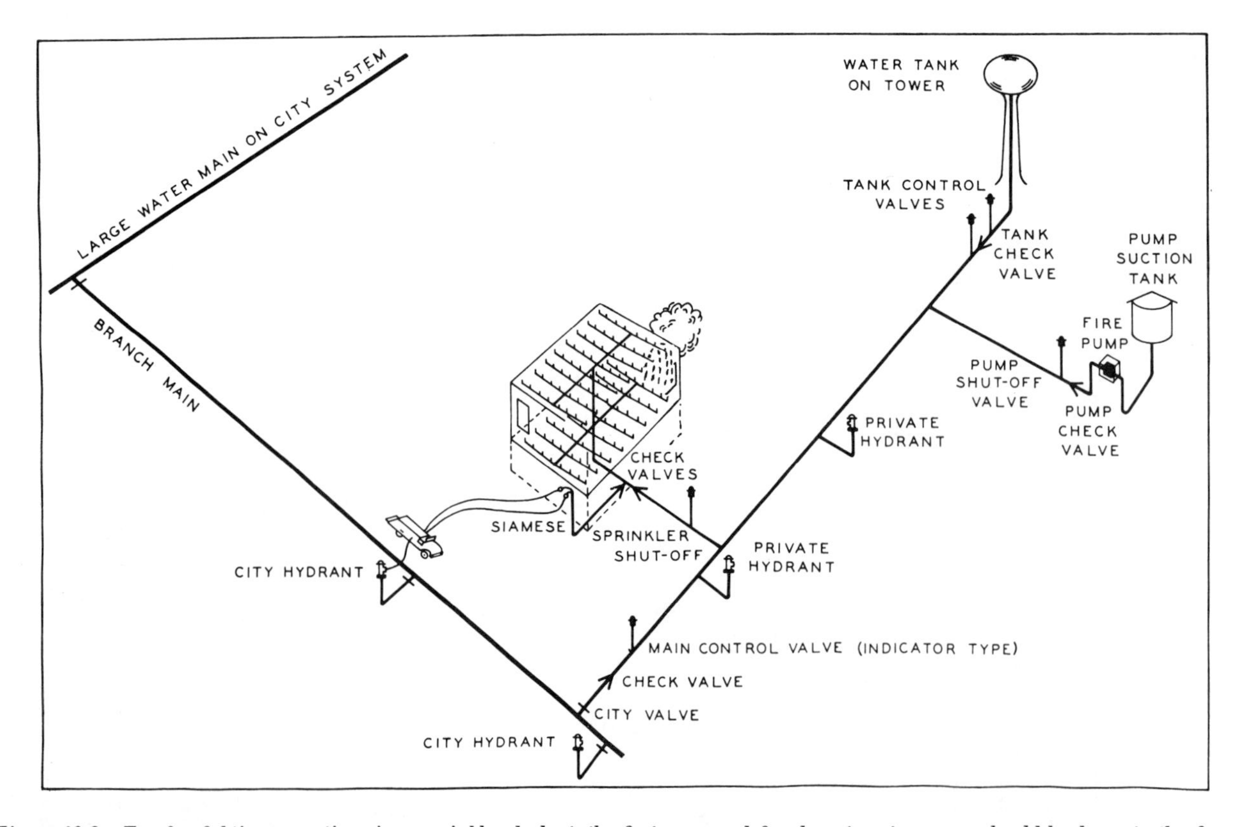

Figure 10-2. For fire fighting operations in a sprinklered plant, the first or second fire department pumper should hook up to the fire department connection.

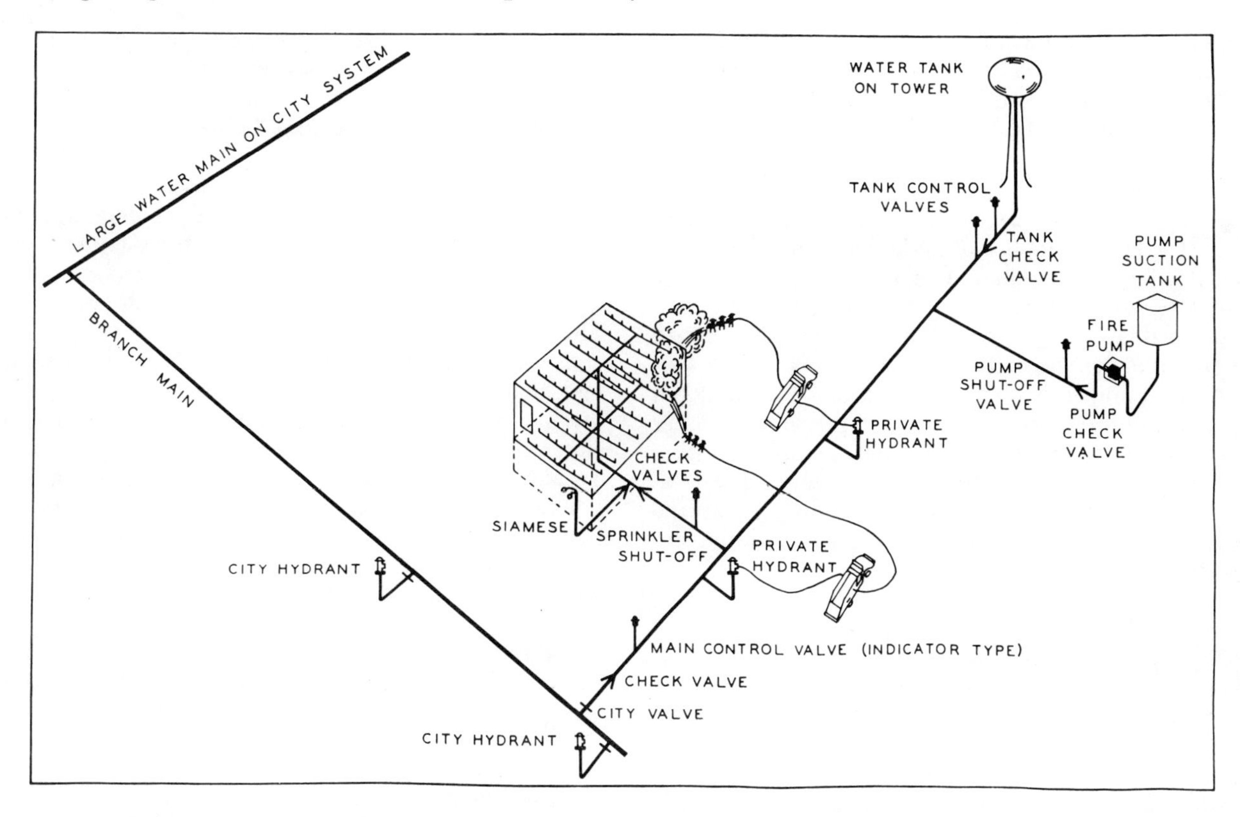

Figure 10-3. For fighting fires in sprinklered buildings, hose lines should not be connected as shown in this illustration because pumpers will take water away from the sprinkler system.

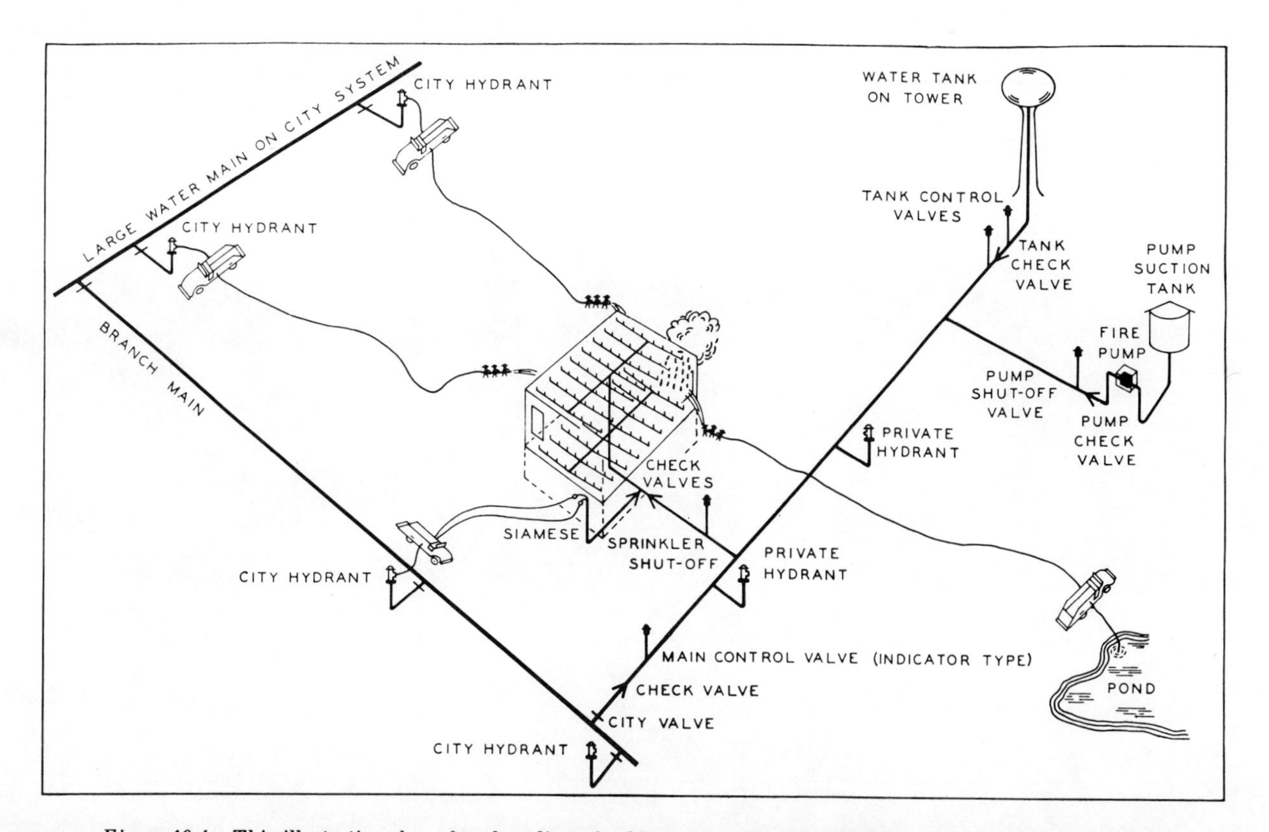

Figure 10-4. This illustration shows how hose lines should be connected when fighting fires in sprinklered buildings.

of the private hydrants. Private hydrants generally are provided for direct hose lines in unsprinklered plants, for piles of lumber or other stock in the yard, or for fires on the outside of buildings.

About ten to twelve sprinklers can be effectively supplied by a single 2½-inch (62-millimeter) pumper line. If a larger number of sprinklers appear to have opened, at least two lines should be connected to the sprinkler siamese and adequate discharge pressure maintained at the pump. It may be satisfactory to provide 100 pounds (689 kiloPascals) pump pressure when the pumper is standing by a small sprinklered fire, but when it appears that an appreciable number of sprinklers have opened, normal pump operating pressures of 150 to 200 pounds (1034 and 1379 kilo-Pascals) should be maintained, depending upon the length of hose line between the pumper and the fire department connection.

Rule 3. Many private water systems are not designed to supply pumpers from hydrants on the private system. Pumpers should not be connected to private hydrants for the purpose of supplying hose streams, unless the system is known to have been designed for this purpose. In addition to connection to city water mains, private sources of water for fire protection may include an elevated tank, a fire pump, or frequently both. These sources are intended only to supply plant fire protection equipment. Suppose pumpers are hooked up to a yard hydrant. Too much water diverted to hose streams may reduce the residual pressure in the water system supplying the sprinklers sufficiently to permit the fire to spread out of control.

Rule 4. Hose streams may be needed to protect exposures or to assist sprinklers at a fire in piled stock or concealed spaces. It is good practice not to take these lines from hydrants on small mains supplying sprinklers but rather to take them from large mains, farther away if necessary, which have sufficient capacity to supply both the hose lines and the sprinklers. The pumpers may take suction from a pond or reservoir, if available. In general, fire departments should find that 60- to 100-gallon-per-minute (227- to 378-liter-per-minute) streams from nozzles capable of either fog or straight stream are adequate for fires in sprinklered buildings. Properties that allow storage practices too severe for sprinkler systems, such as certain types of high piling, cannot expect good fire control even if additional water supplies and large hose stream facilities are provided.

Fighting fires in plants with private fire protection calls for exact knowledge of water sources and their capacities, of piping details, and of the functions of valves and fire department connections, as well as of other details too numerous to be shown in the example in this chapter.

Suggested Reading

NFPA 13E, *Recommendations for Fire Department Operations in Properties Protected by Sprinkler and Standpipe Systems*, 1973, NFPA, Boston.

Chapter 11

VENTILATION and EFFECTIVE ATTACK with HOSE STREAMS

Automatic sprinklers make a significant difference in the situation a fire presents to the fire brigade. In a sprinklered building, the first fire fighting crews must make sure that the water supply is functioning. They may also perform ventilation, salvage, and clean-up operations and restore protection.

However, if hose lines have to be employed in unsprinklered buildings, ventilation and positioning of the attacking crews become complicated problems. In a sprinklered building, ventilation may be much less of a problem, because the fire is usually kept small by the sprinklers. Fire fighting with hose lines may require moving great amounts of water, much more than would be needed in a sprinklered property.

Hose lines have to be used in unsprinklered buildings and on fires in stocks of raw materials or finished products in factory yards. The effective use of hose streams in fire fighting requires a great deal of knowledge about how fire behaves. This chapter attempts to cover the principles involved and the useful practices that members of a fire brigade should follow.

OPEN AND CONFINED FIRES

The Open Fire

The nature and mechanics of fires to be fought can best be understood by starting with a simple outdoor fire like a bonfire. In such a fire, the heat from the burning fuel causes the air immediately above the fire to expand and to become lighter than the surrounding cooler air. The heated air travels upward, carrying with it gases distilled from the burning material and sparks. Heat is also radiated outward in all directions. In fire fighting, the heat that is radiated outward laterally or horizontally presents one of the fire fighting problems. Wind, if any, causes the column of rising air to lean in the direction it is blowing and tends to carry sparks and embers with it.

Near the ground, cooler air moves in to replace the rising heated air. Fire fighters take advantage of this movement of cool clean air in approaching a fire to within attack range. Unless there are unusual wind conditions, the amount of radiated heat generally determines how close hose crews can get to a fire.

The Confined Fire

When a fire occurs inside a building, the same factors are at work as in the open fire. This fire, like the open fire, is free burning until the supply of oxygen inside the building decreases. Fire in a closed building, therefore, gradually builds up heat, which rises to the ceiling or as high in the building as it can go. Heated air develops a small, but significant, difference in pressure between the inside and outside of the building. Near the top of the building, some of the heated air and fire gases will escape through small cracks in the structure. At the same time, cool air will be drawn in around the sides of the building to the extent that it can penetrate small openings.

Such a fire burns freely as long as it has sufficient air. Later it enters a period during which the burning rate slows down because only a limited amount of air can reach the fuel. In an absolutely airtight building, such a fire would conceivably put itself out. At some stage, the burning rate of a confined fire will slow down materially and adjust itself to the amount of fresh air entering the building and the amount of heat and fire gases that escape.

FIRE ATTACK METHODS

There are two general methods of dealing with a fire that is confined in a building — the indirect attack and the direct attack. Fire fighters must know when to use each method.

As long as a fire is confined, its burning rate will be

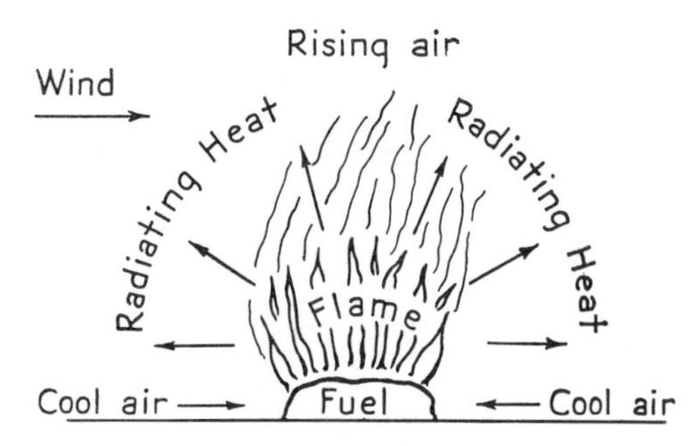

Figure 11-1. Diagrammatic representation of factors at work when a fire is burning in the open.

Figure 11-2. Diagrammatic representation to compare conditions that exist when fire is confined in a building and when it is in the open.

relatively slow, and a relatively high temperature will be built up within the confined space. It is under these conditions that the indirect method, which takes advantage of the hot confined air, is most effective.

The direct attack involves opening the building at some point, so that the normal movement of air by convection will proceed in a controlled manner. This is generally referred to as ventilation. The principle of fire attack using ventilation techniques is relatively simple. It merely means that fire fighters make the fire go in a direction in which it can be kept under control. The idea is to get enough heat and smoke out of the building, so that fire fighters can enter and put water directly on the fire.

Selecting and carrying out these fire attack methods requires some judgment. It is in such matters that the experienced fire fighter has the advantage.

Indirect Attack on Confined Fire

The indirect method does not involve sending fire fighters into the building, room, or confined space. As a practical matter, the method is feasible only when sufficient heat is present in the confined fire area. It would not be employed on a small fire that had not filled the building, room, or space with heat and smoke. The indirect attack consists of directing fog or spray patterns into the upper part of the room or space, where the heat is most intense. Heat converts the fine water droplets to steam, which absorbs heat and displaces oxygen in the vicinity of the fire. Usually, control of the fire is indicated when nearly all of the smoke issuing from the building appears as white clouds of steam.

There are inherent limitations that make the indirect method far from a universal method of fire attack. The first is encountered when the confined heat is not sufficient for adequate steam production. A second limitation is that, in many cases, there are no convenient openings from which to direct the spray into the heated area. Third and most important, the indirect method implies violation of fundamental fire attack operations. Failure to get hose streams into standard operating positions and to properly ventilate a fire area, under the mistaken premise that an indirect attack will solve all problems, can be disastrous. Even if the indirect method of spray application is employed, hose lines should be readied at all operating positions, and ventilation crews should be ready to open up immediately when the order is given. Even when flames are quickly knocked down by the steam generated from a spray nozzle, normally several times as much water must then be applied in direct attack to finish the job. Rarely will an indirect attack extinguish a fire alone.

Theoretically, the indirect attack method can be applied to a confined fire in building spaces of any size. As a practical matter, however, it is most often employed in small rooms or spaces not occupied by humans.

Direct Attack and Ventilation

Fires in buildings can be handled in much the same way as open fires would be handled as long as the fire is small and as long as there is sufficient air available to support combustion. This is the situation that exists when fires are in their incipiency, and it is the situation after a fire has been knocked down and the fire crews are engaged in cleaning up.

The principles of ventilation and direct attack can be understood with a simple example. Consider a small, one-story retail store building. Assume that, about midnight, a fire started in a pile of rubbish, which was to have been discarded the next morning. The fire burned rather freely for a short period of time, and then, due to the lack of air, it began to smolder.

The principal gaseous product of free burning fire is carbon dioxide, a noncombustible gas. When fire smolders, however, the principal gas produced is carbon monoxide, a highly flammable gas. Other gases, mostly hydrocarbon gases, similar to natural gases, are distilled from burning material by the heat.

In our example, the building filled up with a mixture of combustible gases. The fire continued to smolder, supported by what air that did get in, keeping the building full of a mixture of combustible gases that were hot enough to ignite but that lacked enough oxygen to actually start burning.

Figure 11-3. A series of five pictures showing the stages in which fire spreads to involve the contents of a room, Room is one-fifth scale. First picture is taken 10 minutes after a small fire has been started in a corner of the room.

Figure 11-4. Second in a series of five pictures. Model room has fewer combustible surfaces than might be present in a real room, and reduced size makes photographs easier to get. As a result, the development of the fire can be more readily demonstrated. Picture after 12 minutes.

Figure 11-5. Third in a series of five pictures taken after 16 minutes. At this stage of the fire, fire fighters could still enter the room as a dangerous overall temperature has not yet been reached, and there is still air to breathe and to feed the fire. This demonstrates about the latest stage of this particular fire at which direct attack with extinguishers or small hose lines would be possible.

Figure 11-6. Fourth in a series of five pictures after 18 minutes. The steady progress of the fire up to this point has heated the contents so that all combustibles not yet burning are almost at the ignition temperature. At about this stage, further development of the fire becomes very rapid as long as a reasonable supply of air can get in through doors, windows, or cracks.

Figure 11-7. The fifth picture in the series taken after 19 minutes shows the whole room bursting into flames. This is known as flashover. If fog or water is applied at this stage, the turbulence due to flames carries the water particles to all parts of the room. At the same time, these particles are turned into steam, a process that further increases the turbulence and promotes an efficient heat transfer from the burning material to the water. This helps to explain why fog or spray, used in adequate amounts and applied properly, shows such good extinguishing effect. This series of pictures ends with the flashover stage. Temperatures are high and remain high until the fire is extinguished, but the burning will adjust itself to the amount of air filtering in around windows and doors. In a tight room, the fire could actually smother itself.

(Figures 11-3 through 11-7 from "Fire Research 1952"; reproduced by permission of Her Majesty's Stationery Office, Crown copyright reserved.)

Early in the morning, a passerby noticed smoke seeping out of the front of the building and turned in an alarm — a common situation in this type of fire. Responding fire fighters, unable to see the fire, forced an entry into the smoke- and gas-filled building. When the door opened, there was an inrush of air, then an explosion or puff inside the building; the entire interior of the building was in flames. Fire fighters call this phenomenon "backdraft." If any fire fighters were in the doorway or in line with it, the backdraft probably knocked them out into the street.

This illustrates the ventilation of a fire, but it illustrates the wrong kind of ventilation. It is learning about ventilation the hard way, and it illustrates the importance of understanding fire behavior and backdraft in connection with making a direct attack on a fire.

Test fires have been set in small buildings to study the progress of fire in a small room or building and to record how long it takes for dangerous temperatures to develop. The studies have shown that fires started from a small source burn the combustibles in the immediate surroundings. Sooner or later, most of the combustible material in the room will be heated to a temperature above the ignition point and will burst into flame. The ignition temperature is not very high for ordinary combustibles such as wood and paper — around 400 degrees Fahrenheit (204 degrees Celsius). The sudden appearance of flame throughout the entire space is referred to as "flashover." Fire fighters have to develop judgment to tell whether or not a fire in any room or space has reached the flashover stage. The backdraft discussed earlier is simply flashover delayed until enough oxygen is available.

In a building two or more stories high, heated smoke and gases from a small fire will drift through vertical openings, such as open stairways and elevator shafts, and mushroom in the uppermost space. If the fire is allowed to progress, heat and smoke will accumulate and eventually will fill the entire building. The ventilating technique in this more complicated case involves making an opening in the roof, which allows heat and smoke to escape. Fire fighters can then enter the building and find and attack the fire directly.

This is not to suggest that ventilation makes fire fighting easy. Ventilated spaces may be thoroughly uncomfortable and contain enough dangerous gases to require respiratory protective equipment for fire fighters. If fire brigade members are properly trained and equipped, the business of getting in is principally a matter of resolution. Experienced fire fighters expect to get inside and take punishment. Fires are controlled by applying correct methods in a determined and aggressive attack, inside rather than outside.

In cases where fire fighters find evidence of considerable heat, they may attempt an indirect application of water. Where the area is fairly small and a convenient window is available, fire fighters may knock out a pane — but only after a charged line is in position — and spray water fog through the opening. This will allow relatively little air to enter until the fire area has been filled with steam generated by the water spray. As water is converted to steam, its volume expands 1,680 times, helping not only to smother the fire, but also to carry away heat and the products of combustion. For maximum benefit in applying the spray, the nozzle should be rotated so that the spray sweeps the heated area. This method will provide the most rapid heat absorption and steam generation. It is helpful if fire brigade members can be trained in this technique at a fire school.

Where there is evidence of considerable heat and smoke and there is no direct point of access for hose streams (as in an enclosed storage building of considerable area), it may be desirable to place a distributor nozzle through the roof directly over the fire. This acts as a huge sprinkler and will extinguish a considerable body of fire. These nozzles may be placed

Figure 11-8. Fire in a small store or storeroom illustrates the problem of ventilating for fire fighting.

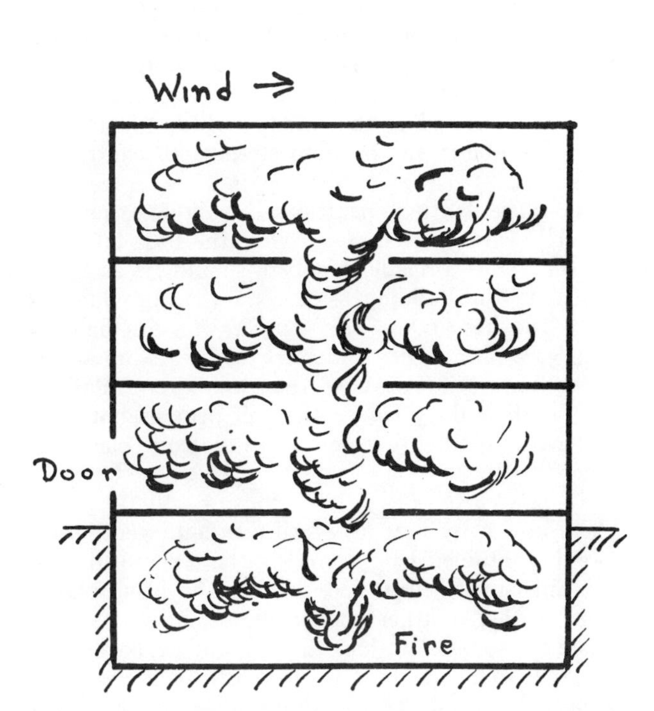

Figure 11-9. A diagram showing how smoke passes up through a building from a basement fire by way of stair or elevator shafts, mushrooming on all floors. At the stage of fire shown, a backdraft or smoke explosion is possible if a considerable amount of air is let in, as by opening a door.

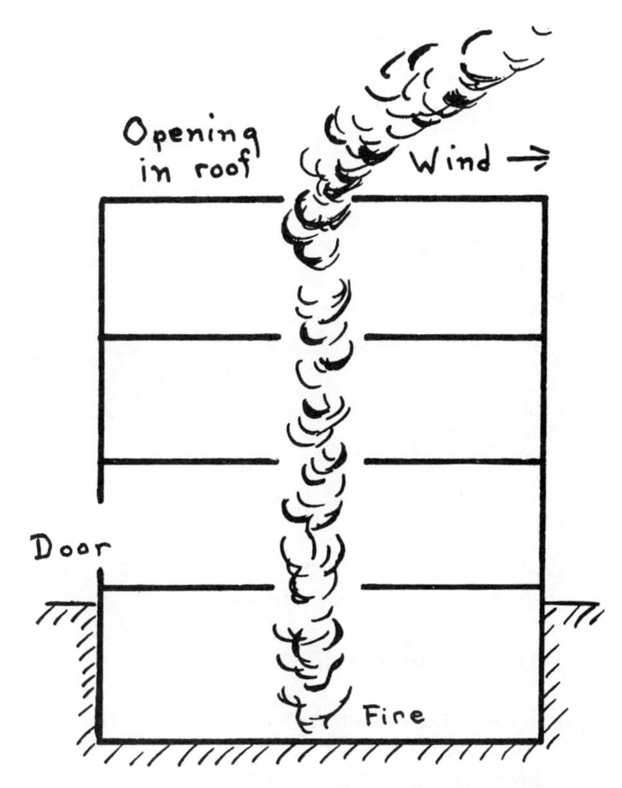

Figure 11-10. A drawing for comparison with Figure 11-9 showing how ventilation makes it possible to clear a building of smoke and heat so that a fire can be located and extinguished. In this case, an opening in the roof carries the smoke up to the top of the building and tends to keep smoke and heat from mushrooming on various floors.

through skylights in sawtooth or monitor type construction. In most of these cases, a distributor made for use with 2½-inch (62-millimeter) hose would be recommended, although successful work has been done with 1½-inch (37-millimeter) distributing nozzles. The large distributor nozzle discharges about 500 gallons (1,893 liters) of spray per minute. It would require pump pressure from a plant fire pump or from an automobile pumping engine to move the required volume of water and have sufficient pressure at the nozzle for good distribution. There are important considerations when using a distributor through a roof. One is to be sure that the roof will support the weight of the fire fighters. Some roofs of industrial buildings are of light unprotected metal construction, which may collapse very quickly when subjected to even moderate heat from a fire. Prefire planning should include noting roof construction. If a roof of any type feels "spongy" to anyone standing upon it, this is a sign that the roof is not safe. In all cases, fire fighters should have a ready path of retreat. But if the roof feels firm and is of substantial construction, it may be assumed to be safe for crews standing on it either to ventilate or to make an opening through which to operate a distributor nozzle.

The second point to consider is that, if there are no openings in the roof, an axe or roof cutter will be needed to cut an opening for the distributor nozzle. If there is a suspended ceiling beneath the roof, a pike pole or plaster hook will be needed to open the ceiling before putting the distributor into play.

Distributor nozzles are not equipped with shutoffs. Therefore, it is standard practice to put a valve in the hose line 50 or 100 feet (15 or 30 meters) back of the nozzle. If the nozzle is supplied by a short line from a pump or hydrant, it may not be necessary to put a valve in the line.

VENTILATION PROCEDURES

Of almost equal importance with the correct application of extinguishing agents is the prompt and effective ventilation of the fire area. Heat confined in a fire area will contribute to the spread of fire. Confined smoke and gases will make the area untenable to fire fighters. Confined smoke, even relatively light smoke, will increase the amount of damage caused by most fires. Therefore, prompt ventilation, coordinated with the application of hose streams or other appropriate extinguishing agents, is a basic part of the fire fighter's job. Where confined heat is intense, pressure builds up, and a serious smoke explosion may occur when additional oxygen reaches the fire. Ventilation crews should

Figure 11-11. For moderate fires that can be approached directly, advance behind a cone of fog directed so as to absorb heat near the ceiling and drive away smoke. In some fires, two small jets are advanced, the second protecting the first, In this direct attack, ventilation on the side away from the fire fighters is desirable so that the heated vapor can be driven outside.

be assigned at all structural fires, and no time should be lost in venting as soon as hose streams are ready (except in cases where the chief decides to apply spray indirectly before opening up).

Where heat threatens to spread a fire laterally under a roof, an adequate opening or openings should be made promptly above the fire. If skylights are available, they can be removed or, if necessary, glass lights can be broken. Some modern industrial buildings have automatic roof vents. These are highly desirable, but in the case of ice or snow, a fire fighter may have to go to the roof to make sure that the needed vents operate properly. Sometimes there are doors giving access to the roof, which can also be opened.

When opening a roof, keep in mind that the fire will tend to travel toward the opening. Therefore, if it is not possible to ventilate directly above the fire, make the openings in the direction of the desired fire travel. Make sure that there are no suspended ceilings or other obstructions that would nullify the ventilation effort. As a general rule, do not direct hose streams into ventilation openings, for such action will tend to drive back the heat and smoke, making it impossible for other fire fighters to work inside the building.

Very often effective and sometimes entirely adequate ventilation can be obtained by opening windows. If the heat is very severe or if the windows are a type that do not open, it may be necessary to break out glass to obtain the desired venting area. Though glass should not be broken needlessly, the prompt breaking of a few dollars worth has saved many thousands of dollars in smoke and fire damage. Every bit of heat that is released from the building not only reduces the amount available to cause damage, but also lessens the amount that must be absorbed by hose streams to extinguish the fire.

Open windows at the top to release the accumulated heat and at the bottom to let fresh air enter and displace the heat and smoke. When there is a breeze, ventilation may be achieved by opening windows on one side of the building to release heat and on the other to admit fresh air. If there is a strong wind, however, the amount of ventilation on the windward side may have to be restricted, so that the fire will not be fanned beyond the ability of the hose streams to control it.

In most fire fighting operations in buildings, ventilation work begins immediately, while the hose lines are being run. Fire brigade members should be assigned to ventilate the building as necessary to provide access for the hose crew. As soon as the fire fighters assigned are inside, they can start opening windows or roof ventilators to provide the desired ventilation. It is also their duty to see that all persons not concerned with fire fighting are removed from places of danger.

There are many fires at which adequate ventilation cannot be obtained by natural drafts. These include smoldering smoky fires, cases where the atmospheric ceiling or other weather phenomenon holds the smoke down, and instances in which the heat and smoke output is too great to be adequately relieved through available openings. In sprinklered buildings, a cool layer caused by sprinkler discharge may develop near the ceiling. If heat has not built up sufficiently to penetrate this layer, roof venting may be ineffective. In such situations, fire fighters use electric or other power-driven smoke ejectors. These are high capacity fans, which can be used to pull smoke through window, roof, or door openings or to bring in needed fresh air.

Smoke ejectors are also useful in removing various gases. For safety's sake, it is desirable to use smoke ejectors that are driven by explosionproof motors, even though there have been few, if any, cases of smoke or gases being ignited by fan motors.

Smoke ejectors have also proven to be of value in salvage operations for removing contaminants from the air during and after fires. In some cases, they are used to distribute deodorizing agents where suitable equipment is provided for this purpose.

The smallest smoke ejector fan recommended for fire service is rated at 5,000 cubic feet per minute (2.4 cubic meters per second). Obviously, where large areas of heavy smoke are involved, this capacity may not be adequate. Accordingly, fire brigades frequently may have to use a number of fans or larger fans to obtain the needed air-moving capacity. In some critical ventilation situations, 20,000 to 30,000 cubic feet per minute (9.4 to 14.2 cubic meters per second) air-moving capacity may be needed. Smoke ejectors must have an adequate electric power supply and outlets.

In some cases, fans may be used to draw the fire in a desired direction, just as with proper natural ventila-

Figure 11-12. First of a series of six pictures demonstrating how a fog jet works to extinguish a fire in a confined room or space. The first view shows fuel burning briskly before the fog jet is applied.

Figure 11-13. Second in a series of six pictures of fog (impinging jets at left) applied to a fire in a confined space. The experimental space is 4½ cubic feet, large enough to demonstrate what happens. First result is that flames are driven out of the simulated windows.

Figure 11-14. Third of a series of six pictures. The fog stream has caused flame in the space to collapse after one-half second.

Figure 11-15. Fourth in a series of six pictures taken after one second's application of a fog jet. Wisps of steam are beginning to appear from the windows of the simulated building.

Figure 11-16. Fifth in a series of six pictures taken 1¼ seconds after the first picture. Continued application of the fog jet shows increasing amount of steam coming from windows.

Figure 11-17. Sixth picture at 3 seconds. The flame has been extinguished, and large clouds of steam (really condensed water vapor) have appeared on the outside of the building. The same sequence takes place when a fog jet is used on a building fire, but this series of pictures shows the extinguishing effect clearly in its various stages.

Figure 11-18. While the movement of air by large blowers ordinarily is not feasible at fires, electric or gasoline driven blowers are used effectively by fire departments to speed up the process of getting smoke and gas out of a building during clean-up operations.

tion, to allow fire fighters to gain access to the seat of the fire with hose streams.

Basements may be vented at sidewalk windows if the windows are located so that smoke and fire are drawn away from a stairway or hatch through which hose lines are introduced. A basement fire may be vented (or vent itself) into the ground floor. In this case, hose lines must be placed to protect combustible stock and equipment and to prevent further extension of fire. In any case, heat and flame will likely belch out of openings made above a fire for ventilation; therefore, it is good practice to have hose lines ready to protect any nearby combustibles. The purpose of the ventilation is to remove the smoke and heat, so no one should be alarmed when the fire is being relieved in this manner. In many cases, it is good practice to vent at one side of the fire and make the attack with hose streams from the other to drive the heat and smoke outside.

Fortunately, modern industrial plant construction favors single-story buildings without basements. This greatly simplifies the fire fighting and ventilation problem. The chief problem is concerned with large areas and high ceilings and the possibility of fire spreading over and under combustible roofs or roofs of light metal trusses, which may collapse quickly from the heat of a fire.

FIGHTING FIRE WITH HOSE STREAMS

The placement of hose streams at a fire can be of utmost importance in limiting the spread of fire and the resulting loss. Hose stream placement is an important responsibility of the fire officer. The person in charge of a single hose line has the responsibility of seeing that the line is placed in the most advantageous position and operated in the most effective manner possible. This means that the stream must be kept moving and directed as necessary to cool and extinguish fire. There is little advantage in continuing to apply water where the fire is already blacked out, except in cases where it is desired to cool some extra hazardous material.

It is not unusual to encounter situations where, at least at the start of operations, there will not be enough streams of water to quickly put out the fire. The chief officer is responsible for placing all available or needed hose lines to best advantage. The fire officer must get lines placed quickly to prevent any further extension of fire while reinforcements are being brought into action. It is here that judgment and skill are most important.

A first principle in placing hose streams is to place them between the fire and additional fuel subject to ignition. This is to prevent further extension of fire before it can be extinguished. If a building is seriously

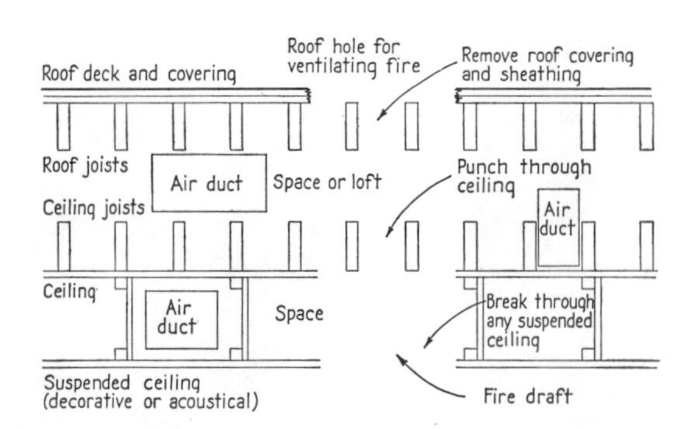

Figure 11-19. A drawing illustrating the problems encountered by fire fighters making ventilating holes through a ceiling below the roof or below an attic space or cockloft. The ceiling as well as the roof deck must be opened. A suspended ceiling, installed for decorative or acoustical purposes, may exist below the regular ceiling, requiring that the opening be made all the way through in order to ventilate the fire. While not designed to illustrate an arrangement of construction that is necessarily typical, the drawing does illustrate the great variety of spaces in which fire may conceal itself at or near the roof of a building. All of these spaces have to be considered in connection with the problem of ventilating for fire fighting purposes.

Figure 11-20. In a building like a pier shed, it may be very important to get a hole cut in the roof, as shown in the upper drawing, if the fire is to be attacked quickly. In a pier shed, storage building, or large manufacturing building, a fire cannot vent itself. It will be forced out through the building itself making access for fire fighting difficult. In a fire, large buildings act like great horizontal flues. Where draft curtains have been provided, holes cut in the roof for ventilation should be on the fire side of a draft curtain, if possible.

involved, the stream is placed to protect any adjoining building or outside storage. In most cases, it is desirable to play most of the water on the main body of the fire, as this will reduce the heat being given off, which can cause the ignition of additional fuel. However, when exposed fuel approaches its ignition temperature and begins to give off smoke, it becomes necessary to wet down that exposed fuel. When this has been done, the stream can again be directed at the main body of fire. In some cases, repeated wetting of exposed fuel may be necessary until the main fire has been extinguished or has burned out. When wetting down exposures, care should be taken not to break window glass with the cold hose stream, as this may allow heat, sparks and fire to enter. If a hose stream is directed at a heated exposure, the water will evaporate and give off light steam. When most of the water has been evaporated more should be applied if needed.

When a hose line is taken into a building, it is important to position it so that it will prevent extension of fire to other parts of the structure. If fire doors are available, they can be closed to delay extension. Often even light partitions and draft curtains can be used in conjunction with a hose stream to prevent further extension of fire, especially if adequate ventilation can be provided. When operating a line inside of a building, it is important that fire fighters have a safe path of retreat. The hose crew should not be in an area where the roof is likely to fall. Where ceiling or roof collapse seems likely, the fire fighters can direct a stream through a doorway or operate from a point where there is adequate structural support. Usually, heavy wooden joist or heavy timber is the safest, as this form of construction does not collapse quickly if the walls are stable and the building is properly constructed. However, everyone should be kept clear of exterior walls once a building is heavily involved. Relatively few fire fighters are injured by falling walls, considering the thousands of fires fought each year. Signs of dangerous masonry walls include loose mortar and smoke or water coming through the brick, or any serious cracks. Sometimes settling noises may give warning of collapse, but these should not be counted upon. Especially dangerous are walls of masonry veneer attached to frame construction. Frequently, these fail without warning. Also dangerous are long masonry walls unsupported by pilasters. Even when a building collapses, the walls usually stand near the corners. The fear of falling walls should not be exaggerated because, in most cases, long before such collapse takes place it is evident that the structure is heavily involved in fire. There is much more danger from falling floors and roofs. There is greater danger of structural failure due to explosions. Plant fire brigade personnel should be trained to recognize any explosion potential in their own plant.

Where a building is more than one story high, the fire fighting problem is considerably more difficult. Heat rises, while smoke and fire spread quickly upward through any available opening and will involve the upper stories unless countermeasures are taken quickly. When a fire reaches a ceiling or roof area, it tends to "mushroom" rather rapidly under the roof, unless the heat is promptly vented. Fire will also spread upward through any open ducts and through concealed spaces. The area above a fire should be checked as soon as it can be made tenable to guard against this vertical spread. A basic rule in assigning crews to positions for controlling a fire is: "Over and under, front and rear." Where this rule is followed, the fire will be boxed and, if adequate extinguishing capacity is at hand, should spread no farther. Because heat rises, normally it is relatively simple to advance hose streams up stairways from below the fire area, provided that sufficient personnel and hose are available to promptly run the line. Where lines are advanced from both front and rear (or from two sides), the fire will be confined between the hose streams.

An important part of the fire officer's job is to provide adequate cooling capacity in hose streams of sufficient reach to cope with the heat of the fire. In relatively

Figure 11-21. When a fire gets possession of the vertical openings, the center of the structure is likely to be burned out. The diagram shows how hose streams penetrate but a few feet from the outside walls of the building.

small areas with low fuel concentrations (or where automatic sprinklers are controlling a fire), streams from 1½-inch (37-millimeter) fire hose may be ample. However, where there are larger areas, as are common in industrial properties, a number of streams from 2½-inch (62-millimeter) hose are needed to prevent the spread of fire. Large area fires and fires involving fuels having heavy heat output may require several thousands of gallons of water per minute to control and extinguish. Large streams may be needed to provide added reach where the fire is too hot or where it is not safe to place nozzle crews close enough so that the fire can be reached with hand lines. A common and popular size of heavy stream nozzle is of adjustable fog or spray type and discharges 500 gallons (1,893 liters) per minute at recommended nozzle pressure of 100 pounds per square inch (689 kiloPascals). This nozzle normally requires two 2½-inch (62-millimeter) hose lines for adequate supply. It is used for large fires involving flammable liquids as well as for ordinary combustibles.

Even larger nozzles may be used to protect industrial hazards. They may be attached to permanent monitor nozzles supplied by plant fire mains or may be mounted on motor fire apparatus. Such nozzles can only be used successfully where sufficient water is available at the needed pressure. The plant fire chief must know how much water is available, because if more streams are operated than the system (whether motor pumpers or fire mains) can supply, the streams will be ineffective. A good rule is to count each 2½-inch (62-millimeter) hose stream at 250 gallons (946 liters) per minute. Two lines siamesed together will give 500 gallons (1,893 liters) per minute. Four 2½-inch (62-millimeter) lines will require about 1,000 gallons (3,785 liters) per minute if standard nozzles are used. If the same water system is used to supply sprinklers, the needs of the sprinkler system must be taken into account. The good fire officer will not use much more water than is needed. In most fires, one or two 1½-inch (37-millimeter) lines or one or two 2½-inch (62-millimeter) lines will be ample. If 2½-inch (62-millimeter) lines are used, once the main body of fire has been knocked down, it is good practice to attach a length or two of 1½-inch (37-millimeter) hose to the nozzle to permit efficient overhauling.

POSITION WORK AT FIRES

Smoke and flame may obscure the true nature and extent of the outbreak. However, the fire brigade's understanding of a few principles applicable in all such instances will do much to reduce fire fighting problems to a routine operation in which skills discussed earlier in this text can be readily and effectively applied in a logical and systematic manner. Fire fighters can do jobs that seem impossible once they know the principles of fire fighting, have practiced basic operations, and are provided with proper clothing to protect them against heat and respiratory equipment to protect them against fire gases.

There are definite tactical positions which experienced fire fighters usually consider essential to control fire in buildings. No. 1 is the possession or control of inside vertical openings (stairways, elevator and other shafts). The rising heat, flame, and smoke from a fire must be denied unrestricted access to these vertical openings if the fire brigade is to retain control and deny the flames the possession of interior means of travel, which also may be necessary for fire fighting and rescue operations. When flames gain possession of these vertical openings, the fire fighters are kept outside, giving flames a chance to spread largely unhindered. Hose streams applied from the outside have limited effectiveness, particularly on upper floors. In such cases, it usually is not long before the roof structure has burned away and the building has become fully involved in flames. Hose streams so applied would be principally to protect exposures endangered by a fire that is out of control.

Where fire fighters control the vertical openings with their hose streams, they can move up under the fire and take advantage of the interior lines to localize and extinguish flame. Some fire fighters can be at work ventilating the building to let out excess heat and smoke, permitting the hose crews to work inside with minimum discomfort. In such operations, all hose crews and other fire fighters maintain a position that affords them a way of escape in event of an unusual and unexpected spread of fire.

Covering the Rear

The fire brigade must always protect exposures and confine the fire so that it cannot extend farther. This is also accomplished by placement of fire fighting crews in a systematic manner to cover essential positions. Most frequently, the No. 1 position is the front entrance of the building, where the line can move in to cover the vertical openings as previously described. Often this line (or lines) can move in to extinguish the fire without further ado. The No. 2 position is usually the rear of the fire area or building. Serious exposure hazards are often at the rear. Sometimes there are persons in the rear to be rescued. The main attack usually will be made from one direction, so that the crews at front and rear will not be driving heat and smoke toward each other. Nevertheless, it is important that both the front and rear positions be covered to prevent extension of the fire.

Getting Over the Fire

The No. 3 position to be considered is a line over the fire, because heat rises and may quickly involve interior exposures in the upper part of the building including the roof structures unless effectively countered by the cooling effect of hose streams. Streams should

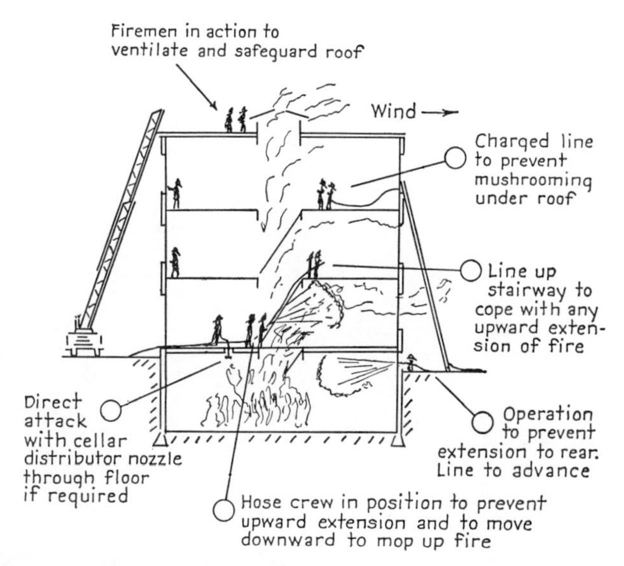

Figure 11-22. A drawing illustrating the principal positions in fire fighting tactics. The drawing shows the relation of ventilation to these tactics.

also be available to prevent extension of fire or damage to exposures when the roof is ventilated. Lines and ladders are usually needed over the fire at both the front and rear of the structure. Where there are no side exposures, such procedures have the fire effectively boxed, ready for the proper lines to move in for the kill. The crew in No. 1 position follows the cool air currents along interior lines advancing under the fire with relatively little discomfort, while the lines over the fire channel and confine the heat.

Side Exposures

Serious side exposures are unprotected windows and other openings through walls or unprotected communicating spaces under the roof or under the building. There must never be any hesitation in opening up to locate and extinguish possible hidden fire that may be spreading through communicating ducts and spaces.

One of the first fire fighting tasks is to make certain that fire doors are properly closed and that any openings through which fire can spread horizontally are safeguarded. Fire fighters should be sent into adjoining buildings to accomplish this. One rule of successful and experienced fire chiefs has been, "Never send one fire fighter alone, send a line," when protecting communicating openings and exposures.

In protecting fire door openings against a severe fire exposure, a rule of very long standing has been, "Never trust the protection of such openings to a single line of hose. Always send two on the chance that one line may fail or prove inadequate." Great care must be taken in opening fire doors, even a small way, to direct hose jets into the fire. At least two lines must be employed if such an opening is made, because the fire may be so intense that, as soon as an opening is made, flame, heat, and smoke may pass through. Fire fighters must be in position to establish absolute control.

The danger of extension of damage must be carefully weighed before any opening is made from an adjoining section of building into the fire area. There are many cases on record where serious losses have occurred, even in sprinklered properties, due to intense exposure fires originating in adjoining areas.

Saving of Life

While it is not possible to lay down hard and fast rules regarding the order in which fire fighting tasks must be carried out, it has long been acknowledged that the saving of life is more important than the protection of exposures and limiting the spread of fire. However, in the vast majority of instances, prompt extinguishment of fire, however threatening and troublesome, is the most effective way of achieving both of these necessary goals.

Where fire streams maintain control of the interior vertical openings, such as stairways, these can be used both for the safe removal of occupants and for entrance by fire fighters searching for possible victims. Where the flames have possession of these avenues of communication, fire fighters and rescue crews are at a very great disadvantage. They have to enter through windows obscured by smoke and heat and frequently blocked by stock in an effort to ascertain that occupants are safe. Then, if victims are found, it is much more difficult to take them down ladders safely than it is to assist them down the stairways.

Accordingly, in most instances the No. 1 fire fighting position is to place a line that gives fire fighters command of the vertical openings for both rescue and fire fighting purposes.

Suggested Reading

Fire Protection Handbook, 14th Edition, 1976, NFPA, Boston.

Kimball, Warren Y., *Fire Attack 1, Command Decisions and Company Operations*, 1966, NFPA, Boston.

Fire Ventilation Practices, 5th Edition, 1970, International Fire Service Training Association, Stillwater, Oklahoma.

Chapter 12

MAKING FORCIBLE ENTRY

Forcible entry means obtaining entrance through locked or secured doors and windows and the opening of roofs, floors, skylights, partitions, and walls by mechanical means to attack a fire or to release pent up heat and smoke.

There are on the market a large variety of forcible entry tools, all similar in general purpose to the ones illustrated. There are also portable, air- or hydraulically-powered forcible entry and cutting tools, which may be available at some plants.

Before beginning forcible entry the fire officer should make certain that it is necessary to do so, because at many fires it is not necessary to force an entrance. Frequently, premises are occupied when fires occur. Often there are unlocked doors and windows, or a patrolman or some other person nearby has the keys. Sometimes certain keys are available to officers of industrial fire brigades. When fires are obviously small, forcible entry may cause unnecessary damage. A ladder raised to a second floor window or fire escape may provide a means of entrance. On the other hand, when an area obviously is seriously involved in fire, forcible entry should be made promptly and vigorously, because doors and windows are already damaged and prompt entrance may prevent further damage to more important structural members and to contents such as expensive machinery and stock.

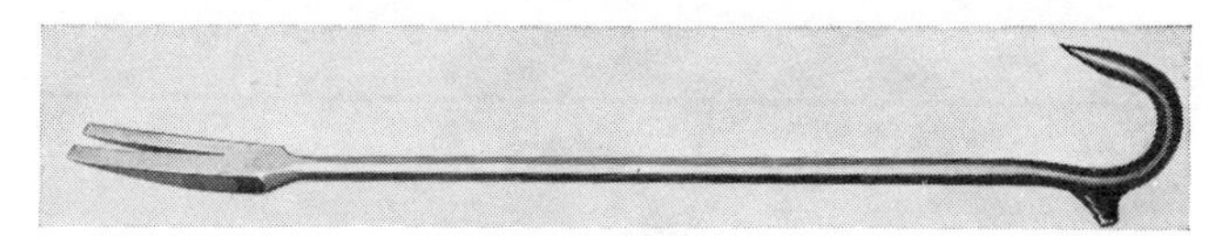

(American-La France-Foamite Corp., Elmira, N.Y.)

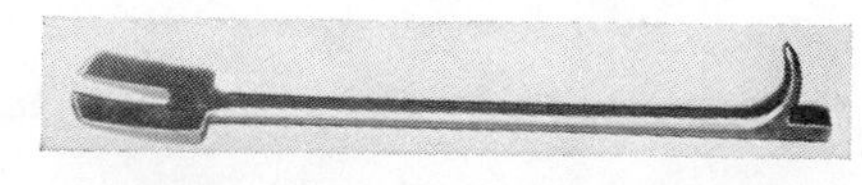

(American-La France-Foamite Corp., Elmira, N.Y.)

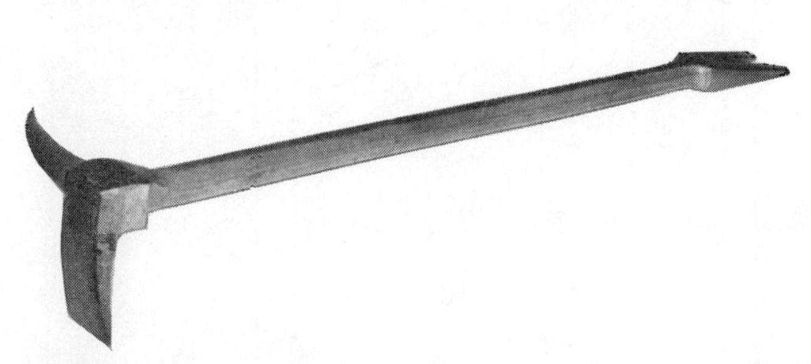

(Halligan Tool Co., New York)

Figure 12-1. Forcible entry tools. Top: claw tool and door opener used to force doors, pry off locks, remove baseboards, pull up planking. Weighs 12½ lbs. is 40 inches long. Center: lightweight, 24-inch claw tool, 6½ lbs. Bottom: Halligan tool, combining hook, claw, punch, lock breaker, and axe. Weight, 8½ lbs.

Breaking a lock or prying open a window in forcible entry in many instances means the destruction of the door or window or other part of the building. In his eagerness to protect property, the fire fighter must not destroy property. As protectors of property, good fire fighters strive to avoid needless damage.

THE AXE AS A CUTTING TOOL

At this point, it is necessary to consider the tool that always has been considered to be the fire fighter's best friend — the pick-headed axe.

The use to which the pick-headed axe can be put at fires are many. To an experienced fire fighter, its

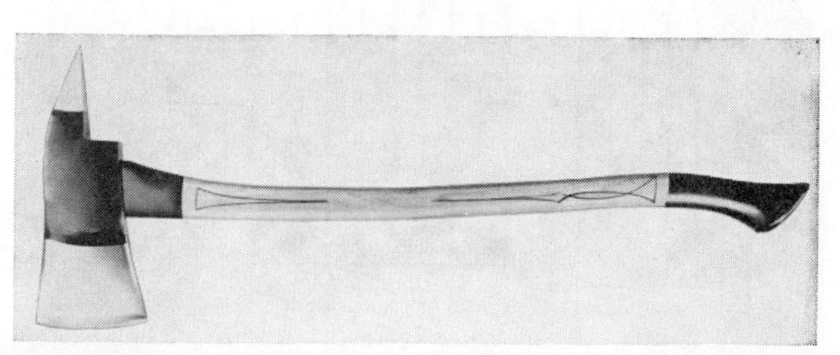

(American-La France-Foamite Corp., Elmira, N.Y.)

Figure 12-2. Fire department-type axe.

adaptability to special uses makes it particularly valuable. Its one great advantage over a common axe is the pick on the back of the head, which is used for prying, ripping, leverage, and other jobs, such as forcing padlocks.

The pick-headed axe has several advantages over the flat-headed axe. It is easier to carry in a belt; it can be used to pull up flooring and roof boards, to open partitions, and for pulling lath and plaster in places inaccessible to pike poles.

If the fire axe were used exclusively for cutting wood free from materials harmful to the blade, a wood chopper's axe would be ideal. If the blade is extremely sharp and the body is ground too thin, pieces of the blade will be broken out when cutting gravel roofs or striking nails and other materials in flooring. If the body is too thick, regardless of the sharpness of the blade, the axe cannot be driven through ordinary flooring.

The recommended body thickness is about ¼ inch (6 millimeters) at ¾ inches (19 millimeters) from the edge, ⅜ inches (10 millimeters) at 1¼ inches (31 millimeters), and ½ inch (13 millimeters) at 2 inches (50 millimeters) from the edge. Measurements are taken at the center of the blade. The standard fire axe weighs 6 pounds (2.7 kilograms), but some favor lighter or heavier axes.

Care must be used in grinding to prevent heating and the consequent softening of the steel. The axe should be ground to preserve the body thickness and not merely to sharpen the blade. After sharpening, the blade may be rubbed lightly over the stone to take off the keen edge and lessen the possibility of cutting fire fighters who may rub against it. An extremely keen edge is not necessary.

The shoulder of the handle is left rather thick to prevent breaking should the axe be driven through a floor unexpectedly. The grip is thin to provide elasticity. and should be smoothed to a polish. Paint is not recommended; varnish or linseed oil may be used to prevent roughness and warping.

CUTTING WITH AN AXE

When cutting with a fire axe, it should not be swung as a wood cutter uses it, but with shorter strokes. Using this method, danger of hitting other men or catching the axe in overhead obstructions is avoided, and the axe is under complete control at all times. When cutting flooring, roofing, or sheathing, the cut should be made at an angle to the grain of the wood, instead of straight across it. About 60° is the correct angle.

Wherever possible, diagonal sheathing should be cut in the direction towards which the sheathing runs. In this manner the chips have a tendency to split out.

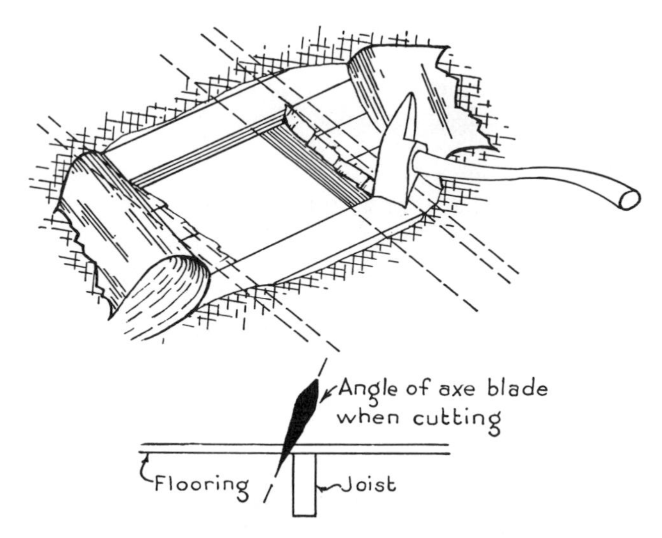

Figure 12-3. Good practice in cutting with an axe.

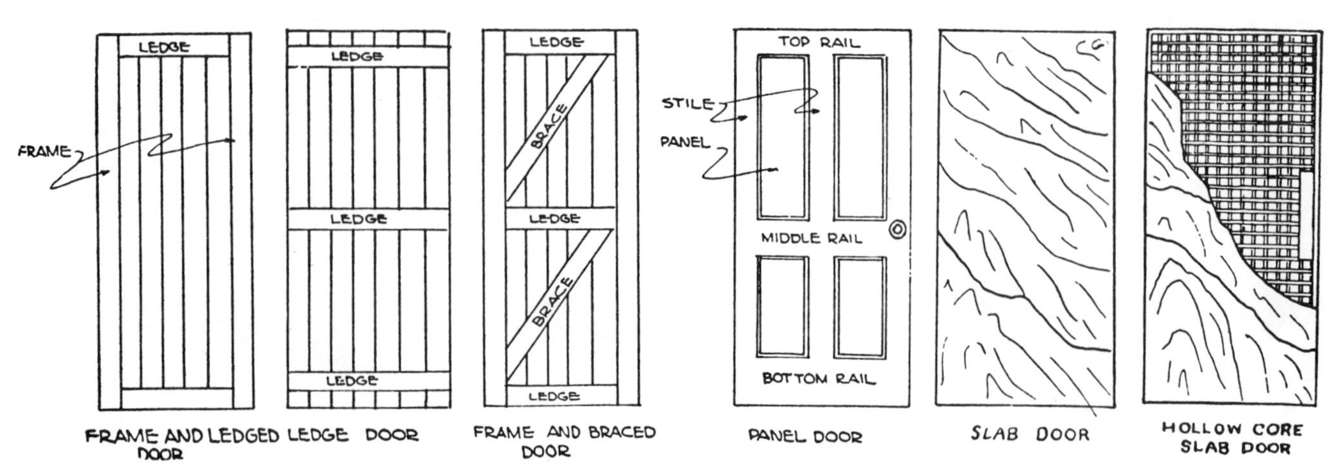

Figure 12-4. Common types of door construction.

If the cutting is done against the sheathing, the axe is apt to bind. If possible, the cutting should be done across the grain of the board instead of straight down. This gives a firmer object to cut against. For the same reason, cutting should be done as close to a joist or stud as possible.

Practice of a standard stroke is desirable to develop the proper technique of using the axe.

When cutting flooring or straight laid sheathing, it may be necessary to cut at a greater slant from the perpendicular than in the case of diagonal sheathing. In cutting through a lath and plaster wall, it is well to remember to cut in a diagonal direction to the grain of the wood instead of straight for the reasons stated.

A fire fighter should be able to cut right- and left-handed. Cutting in difficult corners and under obstructions can be done in a workmanlike manner by brigade members who have been schooled in the proper way to use a fire axe.

(Richards-Wilcox Mfg. Co., Aurora, Ill.)

Figure 12-7. Double sliding-folding doors.

(Richards-Wilcox Mfg. Co., Aurora, Ill.)

Figure 12-5. Double sliding industrial doors.

(Richards-Wilcox Mfg. Co., Aurora, Ill.)

Figure 12-8. Small overhead lift door.

(Richards-Wilcox Mfg. Co., Aurora, Ill.)

Figure 12-6. Double swinging industrial doors.

(Kinnear Mfg. Co., Columbus, Ohio)

Figure 12-9. Large overhead lift door.

(Kinnear Mfg. Co., Columbus, Ohio)

Figure 12-10. Industrial doors — folding overhead lift.

(Kinnear Mfg. Co., Columbus, Ohio)

Figure 12-11. Steel rolling fire doors applied to fire wall openings that are in daily use as service entrances. Rolling fire doors may have various types of manual operation or motor operation.

OPENING DOORS

There are various types of building entrance doors. Ledge doors, sometimes called batten doors, are found in warehouses, storerooms, and barns. They are made of built-up material and must be locked with surface locks, either hasps and padlocks, bolts, or bars.

Panel doors may be of several types, either cross or vertically panelled. The panels are of thin material, dadoed, not glued, into the stiles and rails. Either surface or mortised locks may be used; and hinges may be either full surface, half surface, or hidden butts. The hinges generally have loose pins that can be removed. Front doors of offices often have glass panels.

Slab doors are made of veneered material. A white pine core is generally used, and the veneering may be of any desired hard wood. Slab doors are very solid and not easily sprung.

Industrial doors, such as are used in garages, warehouses and factories, are made sliding, swinging, sliding-folding, overhead lift, overhead folding, or overhead rolling. Double sliding doors are locked together and are very hard to force from outside. Once the lock is released, overhead lift doors are easily operated. These doors are usually locked with sliding bars that must be broken or sprung to release the door.

Overhead rolling doors made of steel offer the greatest resistance of all to forcible entry, because the door cannot be raised except by operating the gear. Prying on the door is likely to spring it so that it cannot be operated. An accepted practice is to install a cast iron plate in the brick wall near the chain. This plate may be shattered with a good blow, thus making it possible to reach the operating mechanism.

The way in which the door is hung and the manner in which it is locked will determine the method of opening. The smaller doors of industrial buildings are either set against stops in the frame or against rabbeted shoulders in the door jamb. With the exception of those in places of public assembly, outside doors usually open in, and the one forcing the door is on the jamb side.

If the door is only stopped in the frame, the stop may be raised with a sharp wedge and the door swung clear of its fastening. However, if the door is set in a rabbeted frame, the opening is not made so easily. Figure 12-16 shows how such a door may be opened, but it is evident that either the jamb will be split away or the lock bolt broken. Figure 12-17 shows how the same door may be opened with the wedge end of a Kelly tool with limited damage. Hydraulically operated spreaders can often spring the frame to release a door.

A door opening out may be opened by inserting a wedge, either a claw tool or the blade of an axe, in the crack between the door and jamb and prying the two apart until the bolt clears.

Double doors may be opened by prying between the doors until the bolt of the active door clears. If an astragal covers the opening, it must be removed before inserting the wedge.

Night latches should yield to the same prying as mortised locks. If they do not, they may be forced off by ramming the door with a heavy object, as they are fastened on the surface of the door with screws. Sometimes this can be done with a man's shoulder.

Overhead lift doors may be forced by prying upward at the bottom of the door with a crowbar, claw tool, or other good prying tool. Once the lock bar is broken the doors should open easily.

Single hinged doors, such as on warehouses and stables, may be locked with a hasp and padlock. If so,

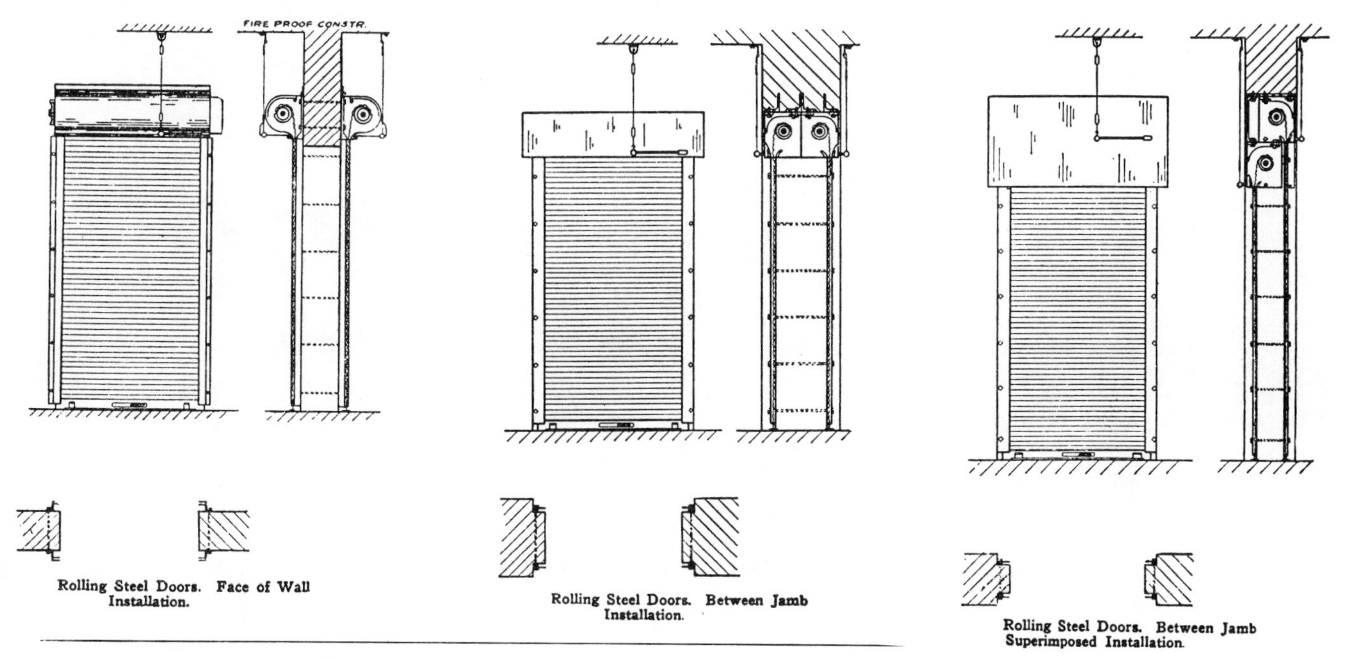

Figure 12-12. Rolling steel doors as fire doors. Three methods of installation.

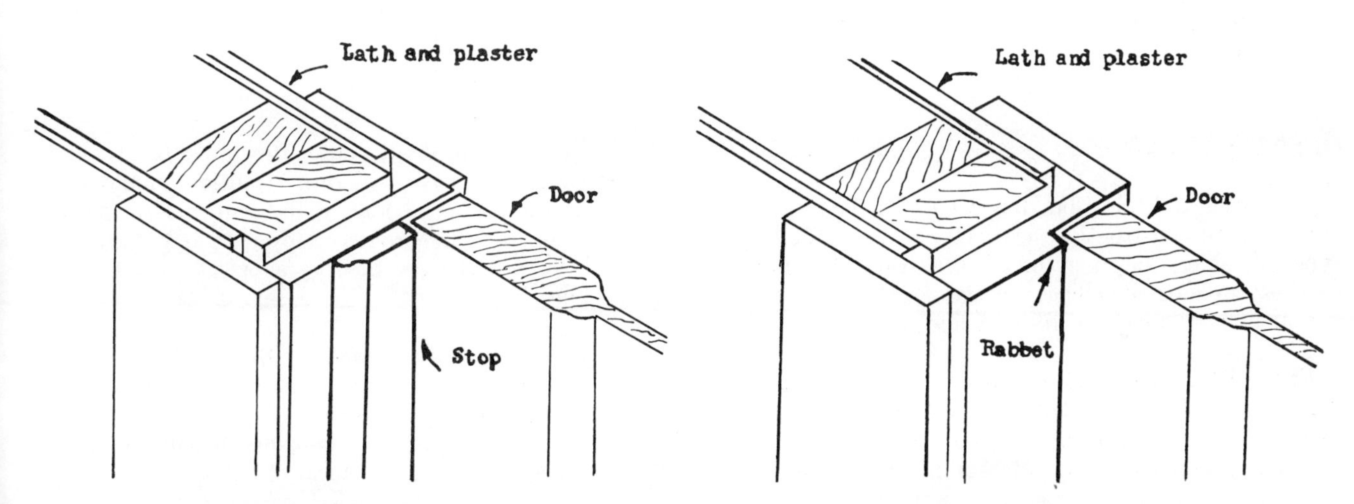

Figure 12-13. Stopped framed door.

Figure 12-14. Rabbeted framed door.

the lock may be opened with a blow from the pick of a fire axe or pried open with the axe pick.

Heavy glass doors, such as those occasionally found at the front entrances of office structures and merchandising areas, are very costly and should not be forced unless absolutely necessary or unless fire has already damaged the door beyond repair. The lock cylinders of such doors can usually be knocked out by a sharp blow with an appropriate tool and this may prevent destruction of the door. Where it is necessary to break the door, it should be struck with sharp blows with the pick of the fire axe but with the fire fighter wearing appropriate fire clothing and standing with face away from the door to avoid possible flying splinters of glass. In opening any door or window of an area seriously involved in fire, there is danger to the operators from flame and smoke when the opening is made. Generally, it is best to work from one side of the window or door opening.

Steel rolling doors often are difficult to open. If a fire crew has to work hose lines through a doorway, it may be wise to prop the door open with a short ladder to prevent the fire line operators from being trapped if fusible links operate the door.

Many times double warehouse doors are secured with a bar dropped into stirrups on the inside of the wall. Forcible entry can be made by battering the door in. Occasionally it may be necessary to make a breach in the wall with a battering ram. The breach should be at a point that will permit slipping the bar from the stirrups.

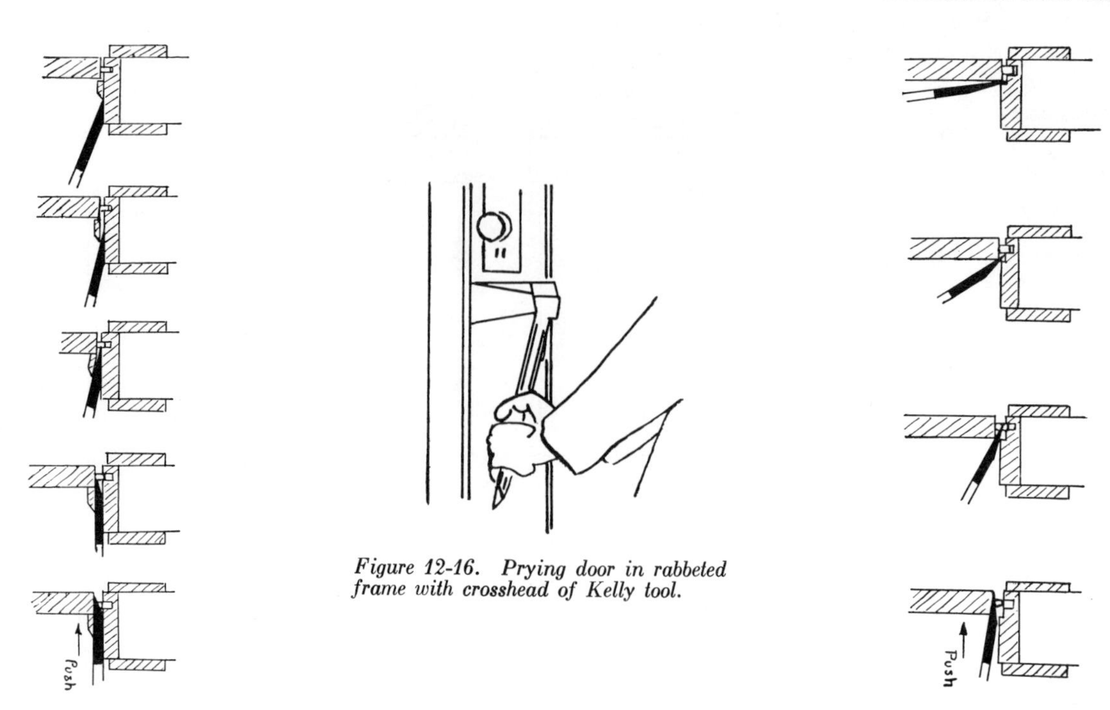

Figure 12-15. Steps in springing stopped framed door with Kelly tool.

Figure 12-16. Prying door in rabbeted frame with crosshead of Kelly tool.

Figure 12-17. Steps in opening rabbeted door with Kelly tool.

OPENING WINDOWS

Industrial and office building windows, if of reasonably modern construction, are likely to be of factory-type. If older, they may be of check rail-type similar to those in most dwellings, or casement-type. Some basement-type windows will be found.

Factory-type windows are made with metal sashes. The sash is often set solidly in the frame, and only a part of it can be opened. The movable part generally is either pivoted at the center or hinged at the top.

Since factory-type windows are glazed with small size glass, it is easier and less destructive to break a glass near the latch and reach in to unfasten it. The rough edges of glass should be cleaned out before the hand is inserted. In case wired glass is encountered, it should be removed from the sash.

The check rail window consists of two sashes that meet horizontally and is the common type used in residences. If the sashes are hung with weights, they will be locked at the center of the check rail, that is, the upper and lower sashes will be locked together. If the window has no weights, the sash will be locked with bolts in the window stiles or a friction lock may be found that secures the sash by pressing against the window jamb. There may be a bottom rail catch. Check rail windows can be opened by prying upward on the lower sash rail. If they are locked on the check rail, the screws of the lock will give and the sashes will separate. If locked with bolts, they must be broken or bent before the sash will rise. Caution should be taken that the prying is done at the center of the sash else the glass may be broken. However, if the check rail latch is on the side, the pry should be made directly under it.

Casement sashes are hinged to the window jambs and meet vertically. Casement windows may be opened much the same as double doors. Generally they are locked securely. Breaking the glass usually is the easiest and most economical form of entry.

Basement sashes, if made of wood, generally are hinged at the top and locked at the bottom. Metal sashes may be hinged either at the bottom or top. Basement windows can be opened with a claw tool much the same as a door in a rabbeted frame. If the prying is done at the center of the lower rail, the lock can be pulled off or sprung.

Remember that prying with a wedge is the principal operation in opening both doors and windows. The fire fighter's axe, the claw tool, or any other wedge-shaped instrument can be used. If the wedge is wide and thin, the entry can be forced with little damage to the building. When necessary, a door can be closed by reversing the operations.

Opening windows to provide ventilation on upper floors of buildings can be done from the inside, possibly after entry has been made from a ladder or fire escape

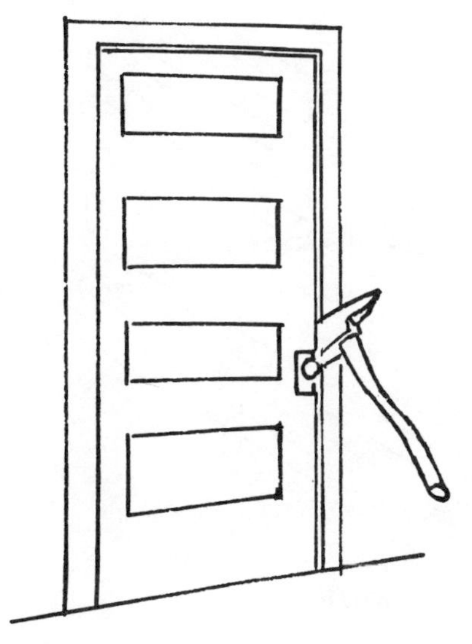

Figure 12-18. Door that opens outward — springing with an axe.

Figure 12-19. Opening a double door.

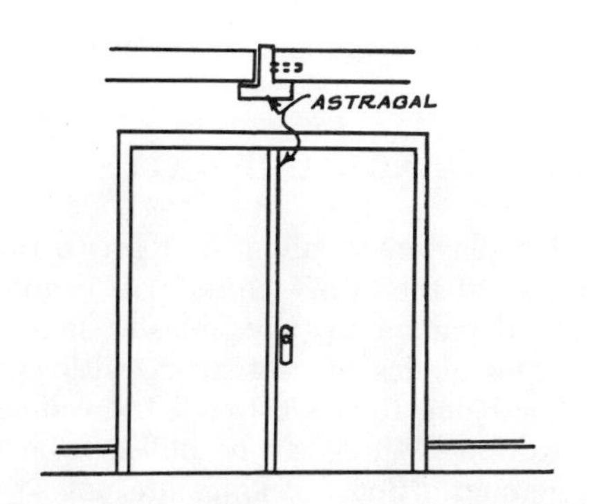

Figure 12-20. Double door showing astragal.

using an axe blade where necessary to gain entrance. If the smoke is too severe for inside work, the chances are that window glass within reach of ladders will have to be broken out with axe or pike pole, although occasionally nearby windows can be forced up or down with a pike pole. Persons opening windows must be careful to see that other fire fighters are not endangered by vented flame, smoke, or falling debris.

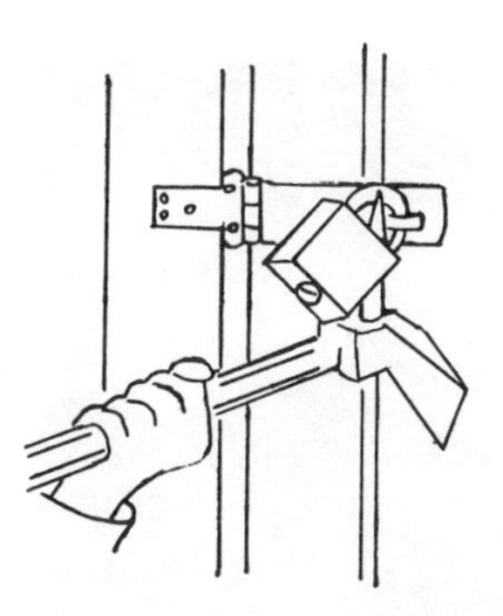

Figure 12-21. Prying off padlock with a tool where lock staple permits.

(Truscan Steel Co., Youngstown, Ohio)

Figure 12-22. Factory-type window.

OPENING ROOFS

In the preplanning of fire fighting, the ability of a roof deck to carry fire fighters should be determined. Some constructions, such as metal joist, collapse suddenly under fire exposures. If, when working on a roof, the roof feels spongy, this is a warning to leave. A shingle roof of any type can be opened by stripping off the shingles and cutting away the sheathing.

Composition roofs are made up of one to six sheets of roofing material, generally asphalt-impregnated felt, nailed to the sheathing and cemented together with asphalt, which is spread over the entire covering to harden as it cools. Some roofs have gravel spread over the hot asphalt. The sheathing consists either of one-inch shiplap laid tight on wood roof joists or of solid concrete (used in newer-type buildings). These roofs require more care in opening, not because they are harder to open, but because they are more difficult to repair. The covering should be ripped and rolled back as shown in Figure 12-3, after which the sheathing may be cut away to make an opening the size desired.

Figure 12-23. Neat cuts in floor or roof sheathing can be made with power saws often available in industrial plants.

Figure 12-24. How to stand and hold axe for cutting. Use a short stroke for safety and accurate cutting.

The roof should be sounded before cutting to determine the location of the joists. Cutting should be close to the joists, making the operation easier and the hole more easily repaired. Openings should be generous. Often it is better practice to pry up entire lengths of two or three planks than to cut a number of small holes.

Metal roofs, generally of tin plate, are sheets crimped and soldered together as one sheet and nailed.

Always work with the wind to your back so you are not affected by gases and fire coming through the opening. Do not remove the planking until the cut has been completed or heat and smoke may make the job impossible.

After a roof has been opened, the ceiling below can be opened by "bumping" it off with a pike pole or any other tool that will reach. Generally a ceiling is not difficult to push off from above.

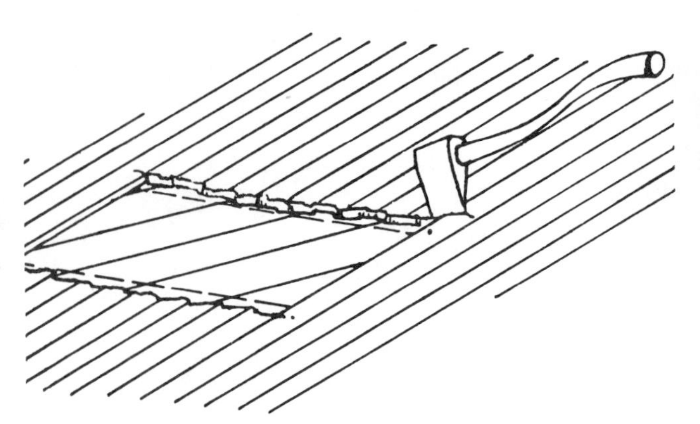

Figure 12-25. Cutting through a double floor.

OPENING FLOORS

Wood floors are usually laid double on 2-inch (50-millimeter) joists set 16 inches (40 centimeters) on centers. The subfloor is laid at a 45-degree angle to the joists, and the top floor, at right angles to the joists.

Floors can be opened much the same as a flat roof, except that, in the case of a double floor, two distinct cutting jobs are required due to the fact that the finished floor and the subfloor run in different directions.

As nearly as possible, the joists should be located by sounding, and both cuts should follow the side of the joist toward the required opening.

The best results with a fire axe are obtained with short, hard strokes. The force-applying hand should be halfway up the handle, and the feet should be spread fairly wide to keep a good balance. Always stand outside the space to be opened. The best results will be obtained by diagonal sloping cuts.

OPENING CEILINGS AND WALLS

To open a plastered ceiling, first break the plaster, and then pull off the lath. A pike pole is good for this job. Some departments have plaster hooks for this purpose. The blades are set apart with springs and expand after being forced through the ceiling. Metal and composition ceilings can be pulled from the joists in a like manner. Board ceilings are not so easily removed due to the difficulty in stabbing the pole through the ceiling and getting the hook fastened above.

When opening a ceiling, take the following precautions. Do not stand directly beneath the space to be opened. When pulling on the ceiling, pull downward at an angle to prevent ceiling material from dropping on your head. Do not attempt to pull down a ceiling without wearing a solid hat or helmet. You should be near an open door through which you can retreat if necessary. Walls of moderate height can be opened

Figure 12-26. Diagonal sloping cuts give best results.

with an axe or other short forcible entry tool rather than with a plaster hook. However, unless excessive heat, discoloration, or other signs of fire are visible, the experienced fire fighter usually will pry off a base board to investigate before damaging the plaster.

OPENING SKYLIGHTS

The general construction of a skylight is illustrated in Figure 12-27. The metal frame of the light itself slips over the roof-flanged opening. By prying under the edge, you can pull loose the entire skylight and then lift it off. If the skylight cannot be lifted and if time permits, glass may be taken out by removing the metal strips or putty covering the joints. However, glass is much cheaper to replace than the roof structure that may become involved through failure to vent the fire, so, as a last resort, it may be necessary to break panes in the skylight to achieve ventilation.

BREAKING GLASS

Glass in either a door or a window can be broken easily by using the flat side of a fire axe. When you break glass, stand to one side and strike the upper part

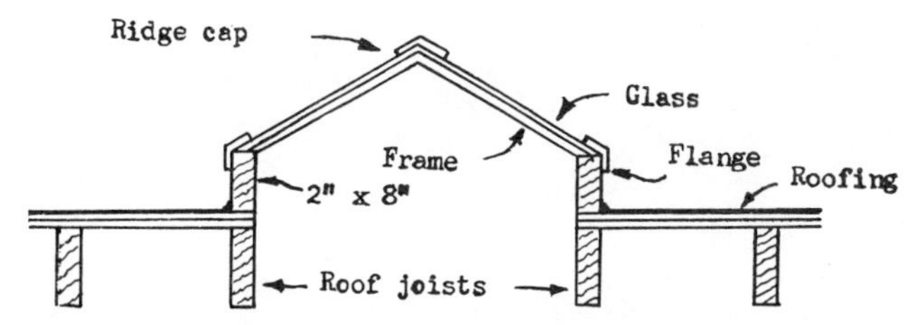

Figure 12-27. Section of typical skylight.

of the glass first. In this way, broken glass cannot slide down the axe handle and cut your hands.

After the glass has been broken out, remove all jagged pieces from the sash. By removing the broken glass, you will prevent whoever goes through the sash from being cut and prevent injury to hose, ropes, or other material that may be passed through the opening.

CARRYING FORCIBLE ENTRY TOOLS

The efficient fire fighter is the one who can work rapidly and safely at all times. To do so, definite practices in carrying tools are as important as knowing how to use them. Never carry tools with sharp hooks or edges on the shoulder. If you should lose your footing, you could injure yourself or someone else. Protection against cuts or abrasions from tools is another advantage of wearing good protective clothing, including gloves, while fighting fires.

Carry any tool with a hook, such as a claw tool, at your side with the hook forward. A pike pole may be shielded by placing a piece of old garden hose over the points, which should be kept shielded until the pole is where it is to be used.

Suggested Reading

NFPA 80, *Standard for Fire Doors and Windows*, 1975, NFPA, Boston.

Forcible Entry, Rope and Portable Extinguisher Practices, 5th Edition, 1972, International Fire Service Training Association, Stillwater, Oklahoma.

Chapter 13

HANDLING ROPE

Because rope is an invaluable tool, fire brigade members should be familiar with some of its uses. A hand line for hoisting and anchoring should be either Manila rope, at least ⅝ inch (16 millimeters) in diameter, or marine grade polyester fiber rope, at least ½ inch (13 millimeters) in diameter. It is generally carried in 100-foot (30-meter) lengths, except where tall buildings make 125- and 150-foot (38- and 46-meter) lengths desirable. Some ¾-inch (19-millimeter) Manila or ⅝-inch (16-millimeter) polyester fiber rope may be carried for possible rescue work.

The hand line should have an eye splice in one end as shown in Figure 13-1. The eye splice avoids the use of knots.

Figure 13-1. Eye splice.

COILING A LINE

For quick use, the hand line must be coiled, so that it will feed out and not tangle when dropped from the top of a building. Two methods of coiling a line are described here.

The simplest way to coil a line for fire service use is in a canvas bag of suitable size. The rope is simply coiled in the bag with an eye splice left outside. When the bag is dropped, the rope is held at the eye splice, and the rope pays out straight as the bag descends. The bag should have a drawstring and shoulder straps for carrying. Sometimes it has an exterior pouch to carry a hose roller.

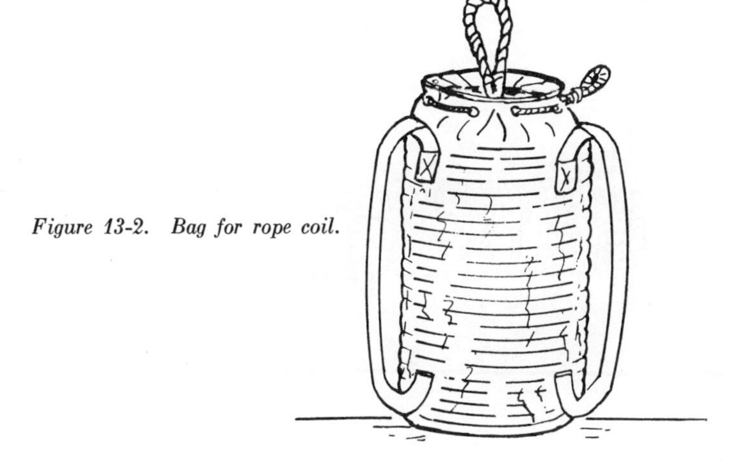

Figure 13-2. Bag for rope coil.

A way of making a coil is to use equipment illustrated in Figure 13-3. The rope roller shown is made from iron pipe. The threads of the straight sides of the tees are reamed away, so that the tees will slip over the pipe forming the axis of the roller. Both tees and the pipe have holes drilled, so that, when the roller is assembled, the vertical bars may be held in place with cotter pins or keys. The right-hand end of the bar, as illustrated, is put horizontally in a hole in a work bench or wooden horse. To start the coil, wind rope around the standards (Figure 13-4).

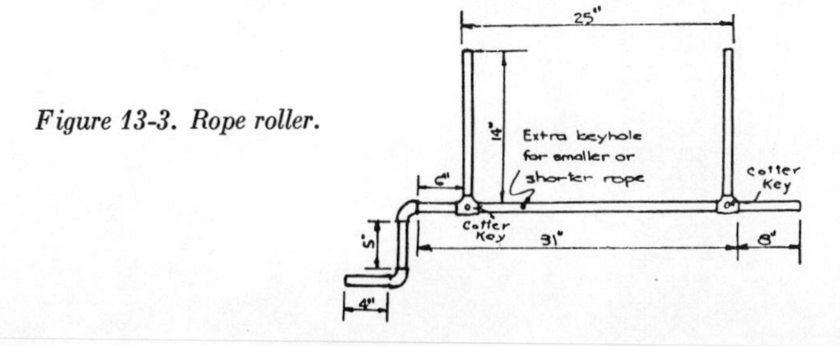

Figure 13-3. Rope roller.

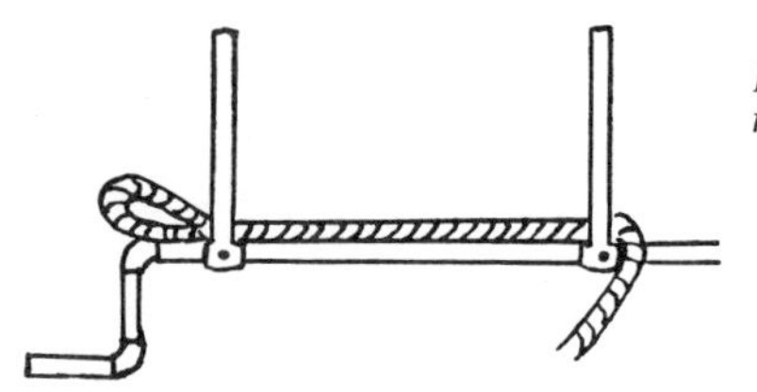

Figure 13-4. Starting a coil of rope.

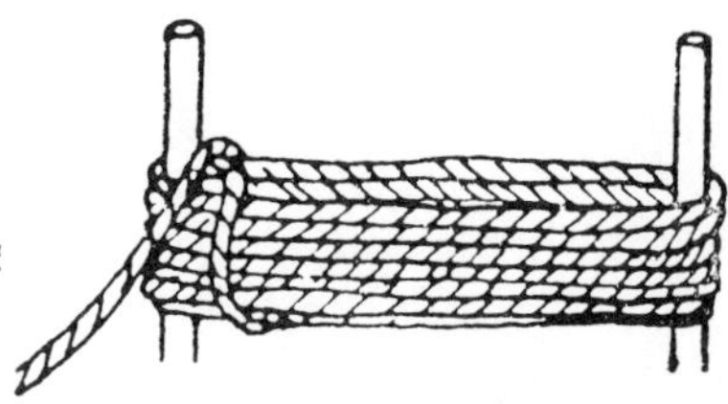

Figure 13-5. Beginning the last course.

Practice will tell when enough rope has been wound around the uprights (Figure 13-5). Then start to turn the rest of the rope over the rope coiled on the uprights. The crank allows the whole assembly to be turned rolling a second coil crosswise over the first one.

When the rope has been coiled all the way horizontally (Figure 13-6), pull the cotter pins and slip the coil from the pipes. Fold the free end and slip the loop through the coil (Figure 13-7). Slip the free end through the opposite end of the coil and through the loop. Pull the loop tight and the coil is finished (Figure 13-8).

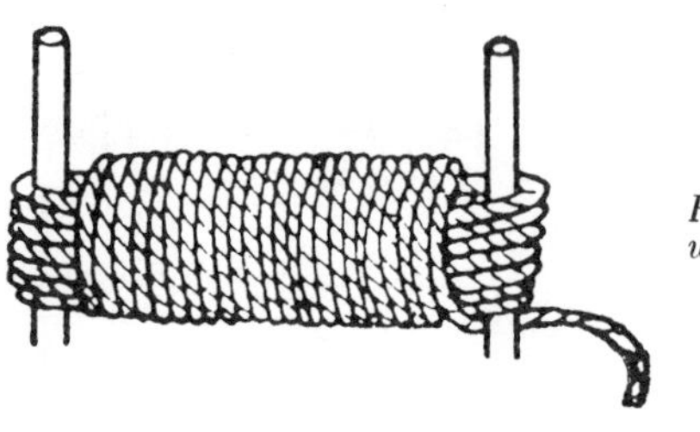

Figure 13-6. Horizontal wind.

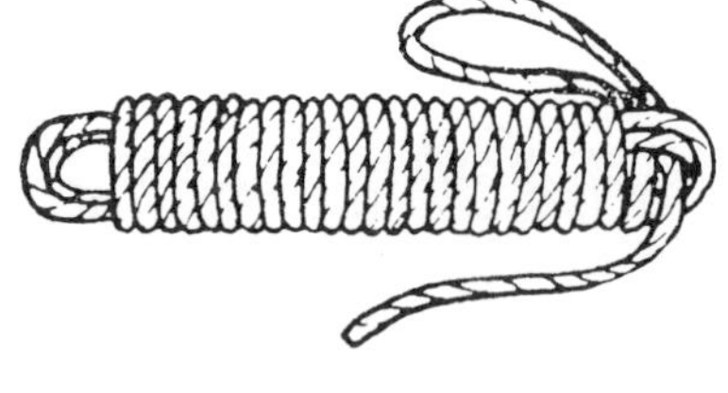

Figure 13-7. Start loop.

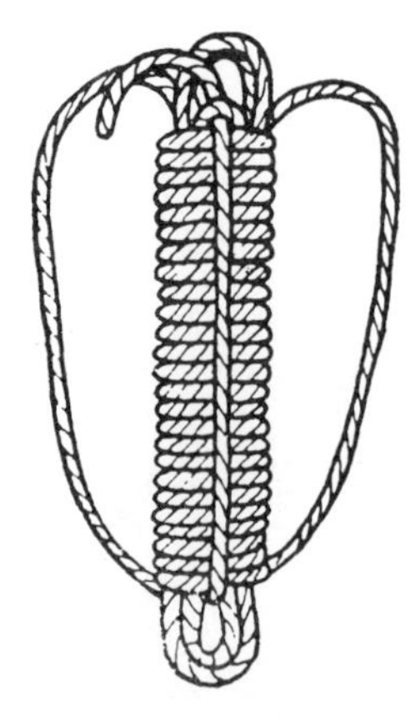

Figure 13-8. Finished coil.

The loops should be large enough for loading the coil on the shoulders. They may be adjusted by pulling the loose end of the rope through the lock loop.

To use the coil, grasp the eye with one hand and pull the loose end from the inside of the coil. Then drop the coil while holding the eye.

TYING KNOTS AND HITCHES

Entire manuals have been devoted to various knots, but most of these are not essential to the fire fighter, who is expected to know a few essential knots and hitches well enough to make them in the dark. First, each knot should serve the purpose required and hold securely until it is untied. Secondly, it should be possible to tie and untie it in the dark, and it should be recognizable by others in the dark. In each case the simplest knots satisfactory for the purpose should be employed.

The hand line is used frequently for hoisting devices or equipment. A knot or combination of knots is needed to tie the tool securely to the line. The clove hitch (Figure 13-9) will hold securely and not slip.

In hoisting a variety of tools, such as a pike pole or fire extinguisher, the clove hitch is tied near the bottom or end of the device, then a half hitch or two are thrown around the other end to serve as a safety and to ensure that the device remains upright while being hoisted or lowered. A half hitch is simply a loop in the rope.

The clove hitch and half hitch are used in hoisting and lowering hose lines. However, it is important that the ties be made as shown in order to avoid damage to the coupling and nozzle. When the hose is being hoisted, the nozzle and coupling must be pointed downward. When the line is being lowered, the reverse is true.

Figure 13-9. Clove hitch.

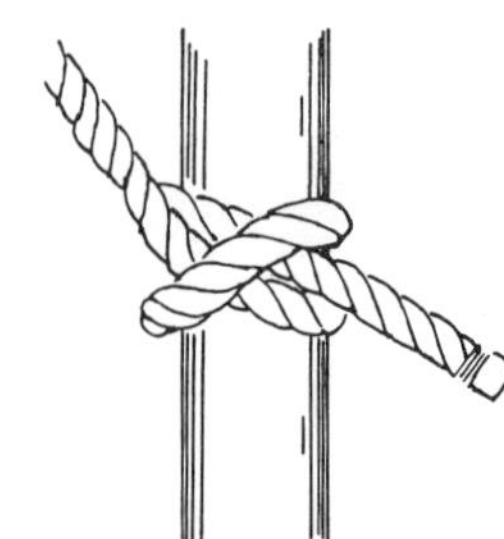

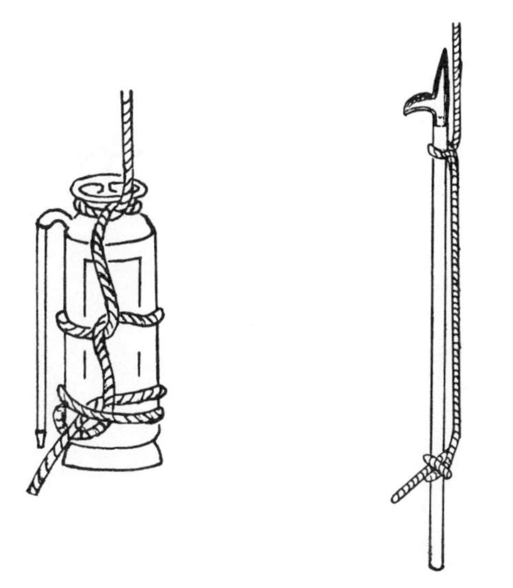

Figure 13-10. Hoisting fire fighting tools using clove hitch and one or more half hitches.

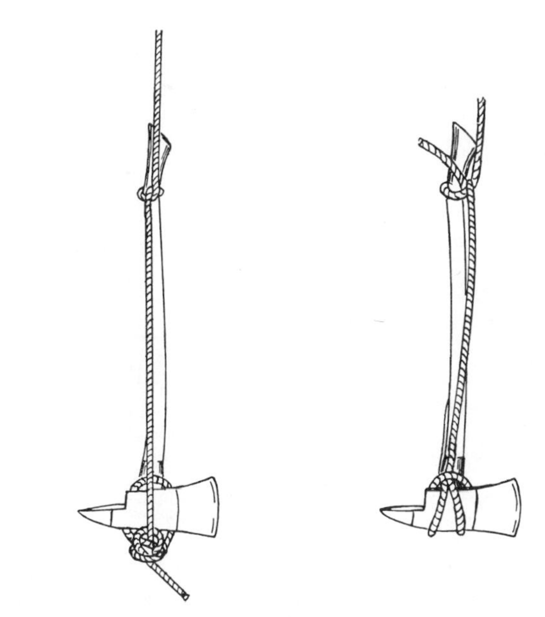

Figure 13-12. Hitches for hoisting axes. Left, bowline and half hitch; right, double loop crossed with half hitch.

The bowline is one of the most useful knots in the fire service. When it is necessary to make a loop in the end of a line, the bowline is used because it will not slip and yet is easy to untie. Due to its shape, the axe is difficult to secure with a clove hitch, but it can be readily hoisted by means of a bowline and a half hitch (Figure 13-12). The loop of the bowline is slipped down the handle. The standing line is then brought around the axe head and secured to the handle by a half hitch. Also shown is a double loop hitch for hoisting an axe. The loose end of the line in the double hitch is secured by the half hitch at the end of the handle.

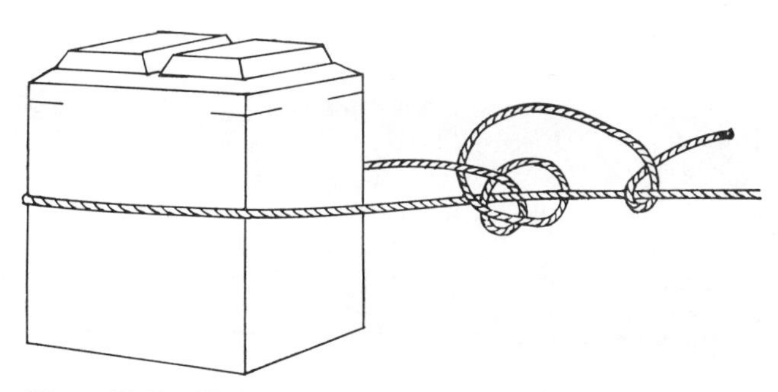

Figure 13-13. Variety of chimney hitch tied loosely to show detail.

The chimney hitch is a good knot to use to anchor to some solid object, such as a chimney or vent pipe. It will not slip and can be easily untied.

If it is necessary to tighten rope, as when stretching guard lines, the half sheep shank can be used. The free end is pulled until the rope is taut, then secured to keep it so.

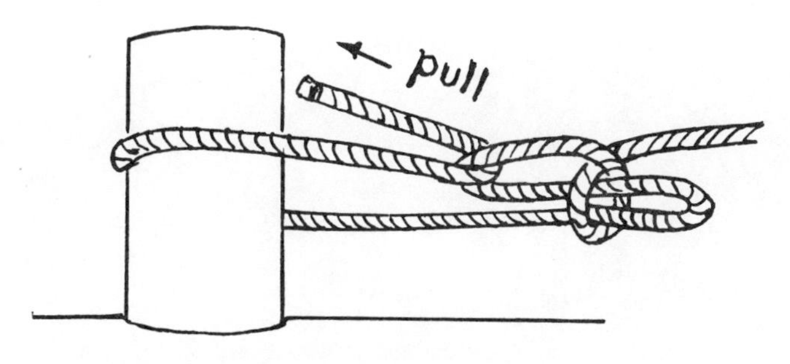

Figure 13-14. Half sheep shank.

The bowline-on-a-bight (Figure 13-15) is a dependable and easily tied rescue knot. To tie a bowline-on-a-bight, double the rope back on itself and tie a simple overhand knot with a loop hanging down as shown. Holding the knot in the left hand, pick up the center of the loop, and place it under the left thumb, forming two smaller loops. Pull on the loops indicated, and the knot is formed as shown in the third step.

Figure 13-11. The bowline.

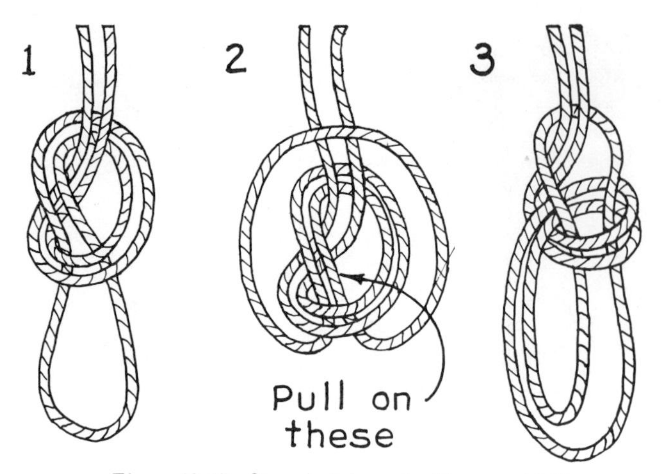

Figure 13-15. Steps in tying a bowline-on-a-bight.

Figure 13-17. Square knot.

To tie ends of rope together, and in bandage work, the square knot (Figure 13-17) can be used.

If the ends of the rope are not the same size, the becket bend (Figure 13-18) can be used.

Two methods of hoisting ladders are shown. A bowline loop is used in one illustrated method. The loop is placed through the ladder one-third of the distance from the top (Figure 13-21). Then it is taken to the top of the ladder (Figure 13-22). The loop is placed over the top and slid down the ladder. The standing line is pulled taut to hoist (Figure 13-23). In this method, the loop and bowline can be tied directly around the ladder one-third of the distance from the top, if preferred.

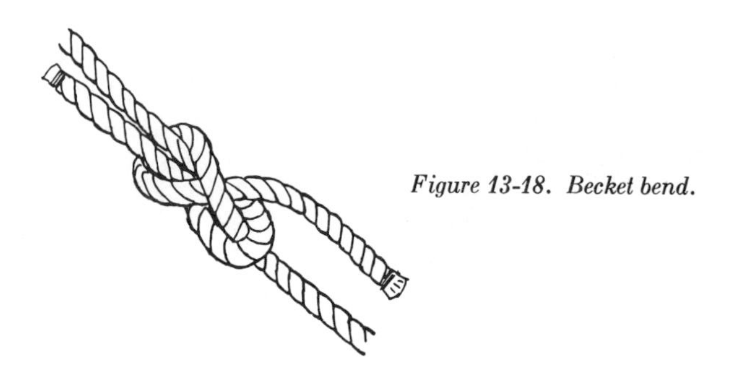

Figure 13-18. Becket bend.

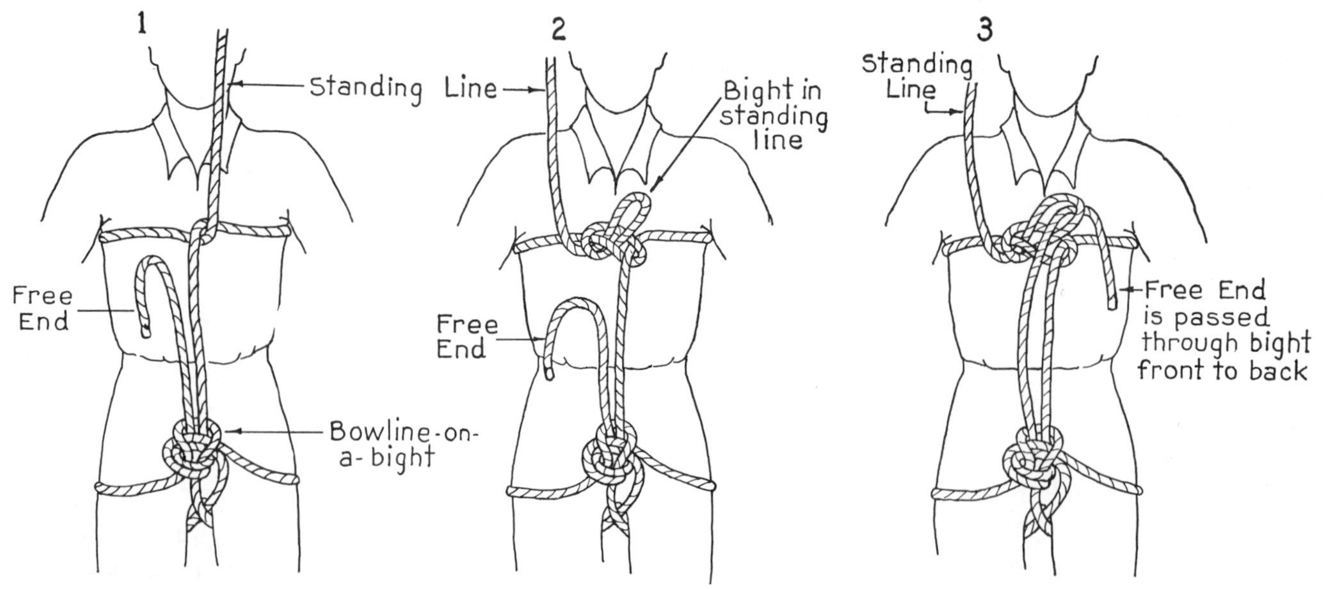

Figure 13-16. Rescue tie for hoisting or lowering trapped persons using bowline-on-a-bight as shown. Two loops of bowline-on-a-bight go around patient's legs and a half hitch is taken under the arms with standing line. A bight is taken in standing line in front of chest and free end of rope is passed through bight from front to back. Pulling on standing rope secures tie.

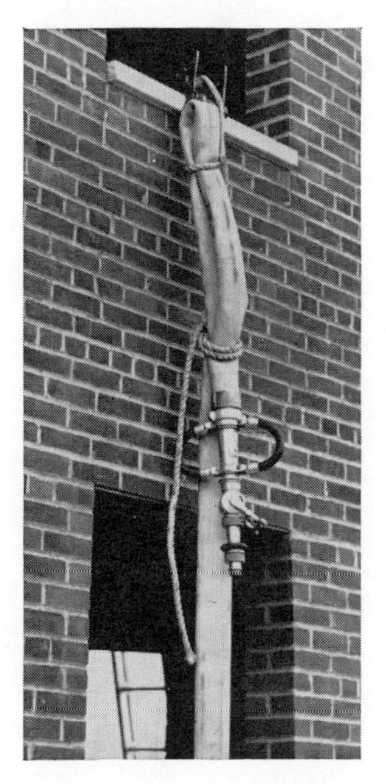

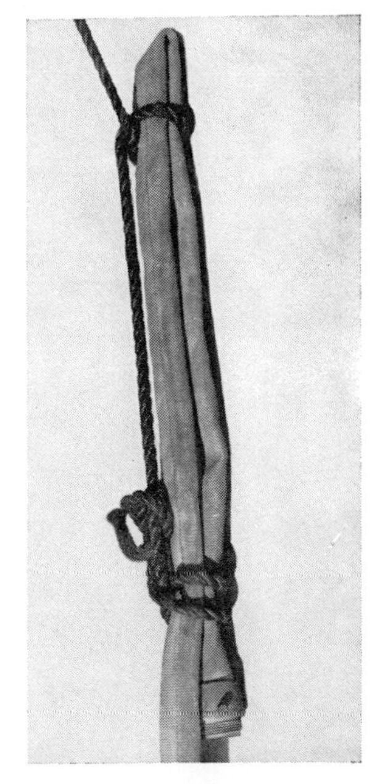

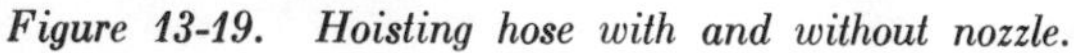

Figure 13-19. Hoisting hose with and without nozzle.

Figure 13-20. Lowering hose with and without nozzle.

Figure 13-21. *Figure 13-22.* *Figure 13-23.*

Bowline hitch for hoisting ladder.

In another method, a clove hitch on one beam may be used to hoist a short ladder (Figure 13-24). The end of the rope is threaded through the ladder one-third of the way from the top. A clove hitch and safety are put on the beam between the first and second rungs at the bottom of the ladder, and the ladder is ready for hoisting. If preferred, the tie may be made to the beam, one rung below the point where the rope is passed through the ladder.

USING ROPE HOSE TOOLS

Rope hose tools can be made with a single line 3 or more feet (1 or more meters) long (Figure 13-25). Both varieties are made of Manila rope, ⅝-inch (16-millimeter) diameter, and have a hook made of ½-inch (13-millimeter) cold rolled steel. When the rope tool consists of a single line, either a metal ring or an eye splice is placed in the end opposite the hook. Before making a splice for use with the rope loop, make four complete twists in the direction opposite the natural twist of the rope to secure a perfect loop when the splice is finished. Take care to remove excess twist when new rope is used.

The rope hose tool should be worn around the outside of the fire coat whenever fire fighters are preparing for duty, so it will be available when and where needed. This tool has many uses in fire department work. It

Figure 13-24. Clove hitch for hoisting ladder.

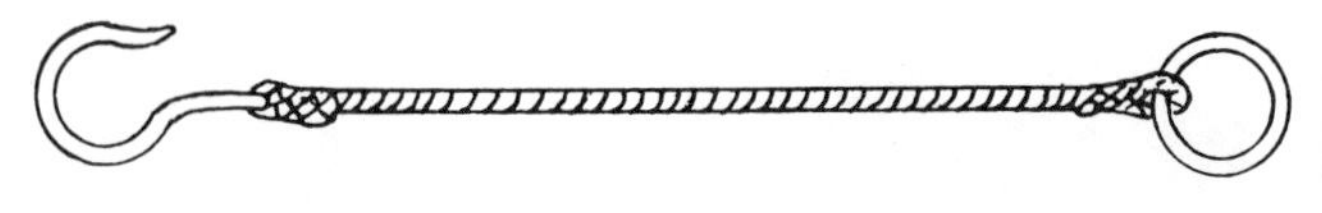

Figure 13-25. Rope hose tool.

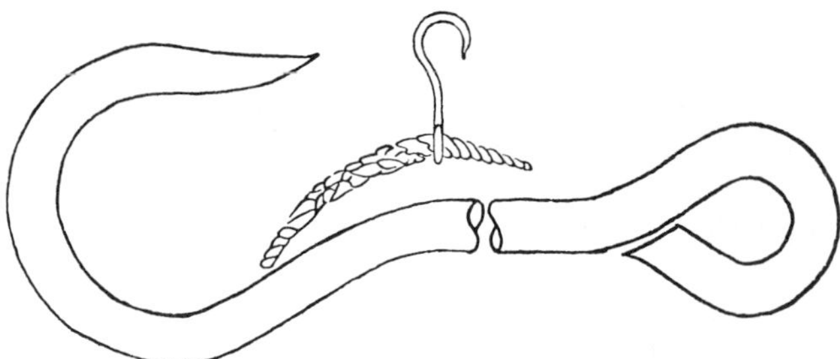

Figure 13-26. Detail of hook and splice for rope hose tool.

can be used as a belt for anchoring one's self to ladders, for securing hose lines to ladders, for securing or dogging ladders to buildings, and for many other uses that will be suggested by experience.

Suggested Reading

Fire Service Training Committee, *Handling Hose and Ladders*, 1969, NFPA, Boston.

Forcible Entry, Rope and Portable Extinguisher Practices, 5th Edition, 1972, International Fire Service Training Association, Stillwater, Oklahoma.

Chapter 14

CARE and INSPECTION of PLANT FIRE PROTECTION EQUIPMENT

Members of the fire brigade are expected to care for equipment assigned to them individually and to assist in caring for equipment assigned to the brigade. Sometimes they participate in various routines established for daily and other periodic inspections of equipment provided by the plant for fire protection.

INSPECTIONS

In general, members of the brigade will not have inspection duties unless they receive special instruction, but the more they know the better they can help keep equipment in order.

An Inspection Routine

Regular inspections are made to ensure that fire equipment is in place and in good order. They also ensure proper attention to housekeeping throughout the plant. Furthermore, the plant may have special fire hazards, and the inspections will see that they are properly protected — for example, that these hazards are safeguarded in accordance with standards of the National Fire Protection Association.

The fire loss prevention manager should make a list of items that are to be inspected. This list should include the following:

(*a*) Control valves on the fire protection piping.

(*b*) Hydrants.

(*c*) Hose house and fire station equipment.

(*d*) Fire pumps.

(*e*) Water tanks for fire protection.

(*f*) Special types of protection.

In addition to these general items of protective equipment, another list of items to be checked should be made building by building, floor by floor, department by department. For each building and department, the list should include such things as the following:

(*a*) Portable fire extinguishers.

(*b*) Small hose.

(*c*) Fire doors.

(*d*) Special hazards, special types of protection, special routines for fire safety.

The purpose of arranging listings this way is to make sure that nothing is overlooked in the course of the inspection. Certain checkups should be made daily in each department.

The person performing the function of fire loss prevention manager for the property should specify what items are to be inspected, who shall make the inspection, and how often inspections shall be made. For most situations, weekly inspections are appropriate. The fire loss prevention manager should specify the reports to be made on inspections and furnish suitable report forms to be used. A routine should be established for review of such reports by each manager concerned.

Inspecting Valves

Most properties where the fire protection has been carefully planned have adequate supplies of water available for fighting fires, either through an automatic sprinkler system or for manual fire fighting. This water comes to the various buildings through a system of piping. It can be readily understood that, if for any reason, either accidentally or deliberately, any of the valves in this system are closed, the water will not be available for fire fighting. This is especially important in the case of valves controlling automatic sprinkler systems, because the prompt, automatic action of the sprinklers is lost if the valve is closed. Even if someone remembers to open a closed valve, the fire may have time to gain considerable headway and cause great damage.

Many plants follow a desirable practice of providing each control valve with a sign indicating what portion of the fire protection piping it controls. In any case, each valve should be given a number so that it can be identified in records kept of its use and of its condition at each periodic inspection.

It is also good practice to have in effect a system so that a responsible executive or loss prevention manager has brought to his attention records of operation of any valve. One system of this sort is described as a "red tag" system. When a sprinkler valve has to be closed for repairs or other reasons, a red tag is hung on it, serving as a constant warning that the valve must be reopened. A stub of the tag is also placed where it will serve as a reminder to a responsible executive or loss prevention manager.

There should also be a system to discourage tampering with valves and to make it possible for personnel assigned to inspection of the valves to be able to readily tell if there has been operation or tampering. An easily instituted arrangement involves the use of a wire and lead seal, similar to freight car seals (Figure 14-1). On the lead seal can be placed an identification of the inspector who last inspected the condition of the valve. Such wires are arranged so that they have enough play to enable the valve to be moved for inspection of its condition where such movement is needed, but not enough to enable the valve to be fully operated without breaking the seal. Where straps or bands with padlocks are provided for a similar purpose, the padlocks should have a brittle shank, which may be broken when a valve has to be operated. For wheel-type valves, a sturdy chain and padlock may be used to secure the valve in the open position.

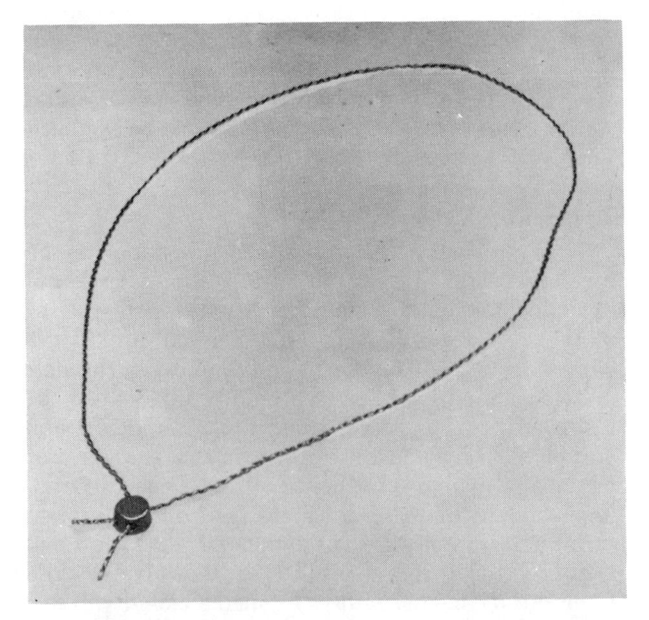

Figure 14-1. A commonly used assembly for sealing valves consists of No. 16 twisted copper wire about 19 inches long and a lead seal that can be slipped over the ends of the wire and squeezed. The lead seal is 5/16 inch in diameter and 5/32 inch thick with two holes for the wire. The wire can be threaded around the valve wheel or handle in such a way as to provide slack enough to try operation of the valve, but to require breaking of the wire or seal if the valve position is changed.

In most properties, valves should be inspected weekly. Occasionally, inspections daily or even more often might be called for in the case of protection to property or part of a property of extreme importance.

Valve inspections should be recorded in reports in which each valve is identified by number and its condition stated.

The record should show that the valve is in the required position. This is usually the open position. Occasionally there is a valve intended to be normally closed. It is important that the inspector be aware of such an exceptional case.

Gate valves do not need to be completely operated as a part of the inspection routine. These valves are left so that normally they are about a quarter turn from the wide open position. In the inspection, they should be turned down a part turn, then up to full open position and then turned down a quarter turn. This slight movement generally assures an inspector that the gate valve and its stem have not separated and that the valve is not jammed in an open position.

The inspection should determine that the stems of outside screw and yoke gate valves are clean. About once a year the stems of all gate valves should be greased. Wrenches for indicator post gate valves and underground gate hub valves should be on the valves or nearby and properly marked.

Listed indicating valves of the butterfly type require only visual inspection of the target mechanism and do not require operation for inspection or annual greasing.

In all cases, the inspection should determine that the valves are accessible. Particularly, the hub box of underground hub gate valves, through which the valve is operated, should not be blocked or covered. Similarly, the access to valve pits should not be blocked; the pits should be clean and clear of water.

Inspecting Hydrants and Hose Houses

Members of the brigade should thoroughly understand how the hydrants in the plant and on nearby streets are assembled, and how they operate. A section of a common frost proof type is shown in Figure 14-2.

Not all hydrants are exactly the same. Some fire fighters training schools have some common types of hydrants with sections cut away so the details of the hydrant mechanism are visible. These are excellent when available for instruction. Sometimes it can be

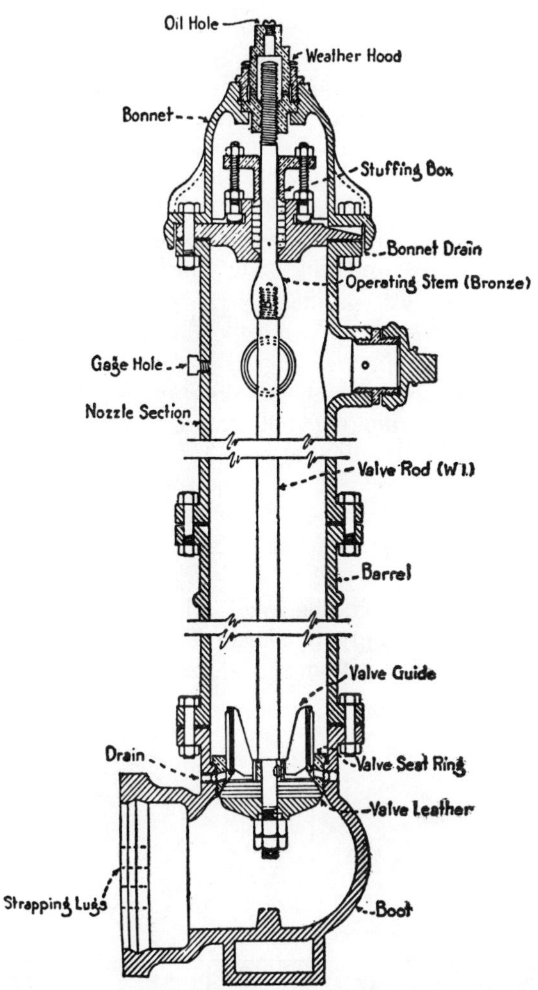

Figure 14-2. Section of common type of hydrant for factory yards. The type shown has the valve below the frost line, hence the description "frost proof." This drawing is for illustrating the features mentioned in the text but does not correctly show details of all hydrants in use. Brigade members should study their local hydrants.

arranged so that all members of the brigade at one time or another can help in hydrant repair work, which may give them an opportunity to see, disassembled, a hydrant of the type in use in the plant. Where neither of these is possible, the instructor can often get a catalogue from the manufacturer of the hydrants in use. This catalogue will have, in most cases, a section drawing of the hydrant in which the various parts are identified. Knowing what the various parts are for helps in understanding why the following listed inspection routines are important.

The first feature of hydrant inspection is to see that the hydrant is undamaged, accessible, and unobstructed. Where there are hose houses, a check should be made to see that hose and equipment are in place. Such inspection should be made daily or weekly.

At least once a year the hydrant should be given a complete operating test by a competent mechanic, at which time necessary repairs are made. Unless all the hydrants at the plant are alike, each hydrant should be marked at that time to show how many turns must be made to open it fully and whether these turns are made to the left (counterclockwise as you look down on the hydrant) or to the right.

At least weekly (daily in freezing weather) inspection should be made to see that the hydrant valve stem is free to move and that there is no water in the barrel.

In most frost-proof hydrant designs, there is a small drain valve at the base of the hydrant near the main valve. When the main valve is closed, there should be no water in the barrel if the drain valve is operating properly. The drain valve is usually a ball valve. In this, pressure inside the hydrant rolls a bronze ball up a slight incline to seat against the discharge opening. When the inside pressure is removed as the hydrant valve is closed, the ball falls away from the opening, allowing the water to drain out of the barrel. If water is found in the barrel, one cause may be failure of this valve to open because the ball is stuck in sediment or held away from the seat by a small obstruction. When the main valve is turned just one or two turns a small amount of water can flow out of this valve. Just "cracking" the main valve this way oftentimes will set up enough flow through the drain valve to blow out sediment and any small obstruction that is preventing the drain valve from closing. If the drain valve is not operating properly and it cannot be cleared by cracking the main valve, in most cases it will be necessary to dig up the hydrant until the drain valve opening is uncovered, and then clear the physical obstruction with a wire or small rod.

A hydrant that has been properly installed must be so located that water from the drain valve can drain away. Usually it is good practice to put about a bushel of small stones under the hydrant and around the drain so water can flow away. Also, the rule should be to always open the hydrant completely. Hence the importance of knowing the number of turns to the full-open position. If hydrants are merely cracked, water flows out the drain valve and, in poor soil drainage, may stand around the hydrant. Then, when the hydrant is closed, the water flows back into the barrel.

Water in the hydrant barrel is sometimes due to ground water rising above the drain opening. While hydrants should be set so as to drain properly, sometimes there are temporary ground water conditions. With these, it may be necessary to plug the drain outlet in freezing weather and pump out the barrel.

Where water gets in the barrel through leaks, a daily routine of pumping out the barrel must be instituted in freezing weather until the hydrant can be repaired, so as to stop leaks past the main valve or through an imperfectly plugged drain valve.

A quick test for water in the barrel is to remove a cap and "sound" by striking the palm of a hand over the hose outlet. Water or ice in the barrel raises the pitch of the note. The change from normal pitch when the hydrant is dry can be detected with a little practice. If a string with a small weight is lowered into the barrel, it will strike ice or come up wet if there is water in the barrel. If ice is solid in the barrel the main valve stem probably will not turn. If it is only slightly bound by ice, tapping the arm of a wrench on the operating nut may release the stem. Moderate blows should be employed to avoid breaking the stem.

Salt or antifreeze solutions are not very good to prevent water in a hydrant barrel from freezing, and their corrosive effect may impair operation of the hydrant.

When inspecting hydrants, look for water on the ground and other evidence of leaks in mains near the hydrants. Where leaks are suspected, there are stethoscope-like listening devices, which may be used to find the leaks.

Inspecting Extinguishers and Other Appliances

The principal maintenance features of portable fire extinguishers have been discussed in Chapter 3. A periodic inspection is generally made to see that all fire extinguisher units are in place. The inspector should note the date of the last inspection of each extinguisher.

Where hand hose sets are provided on standpipes, a check should be made to see that the hose is neatly stacked, clean and dry, that the nozzle is attached, and that the valve has not been leaking.

Inspecting Sprinkler Systems

There are a number of points of importance to be observed in making sure that sprinkler systems are always ready for service. Following are some of the principal items:

(*a*) Check sprinklers to make sure they are the proper type, in good condition, clean, free from corrosion or loading, not painted or whitewashed, and not bent or damaged.

(*b*) Assure that an adequate supply of extra sprinklers is on hand.

(*c*) Make sure that sprinkler piping is in good condition and has not been mechanically injured. See if any hangers are broken and need replacing.

(*d*) Note if any partitions have been erected which require the extension of sprinkler lines.

(*e*) Note cases where stock is piled so closely under sprinklers that the sprinklers cannot readily distribute water over the stock piles.

(*f*) Examine the fire department connection, making sure that caps are in place, threads are in good condition, ball drip or drain is in order, and the check valve is not leaking.

Inspecting Sprinkler Control Valves

Sprinkler control valves are included in the recommended weekly inspection of all valves in the fire protection system. There are a number of additional points to note, however, in connection with the valves and related equipment, such as the following:

(*a*) At the weekly inspection, read the gages showing water and air pressure on the system to make sure that normal pressures are being maintained.

(*b*) At intervals specified by the plant fire loss prevention department, make a water flow test at the main drain valve.

(*c*) At the top of each sprinkler system, there is a small outlet for making a flow test. With this outlet open, the water flow alarm device should operate. This may be observed by a person at the control valve at the sprinkler system riser.

(*d*) Operate test switches on electric alarm devices weekly.

(*e*) Make sure that the small valve or cock controlling the water supply to alarm devices is secured in its normal open position.

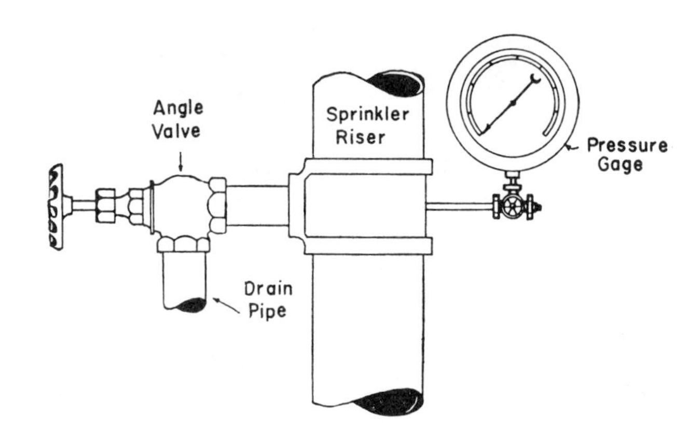

Figure 14-3. Waterflow tests on a wet-pipe sprinkler system are made by opening the drain valve and noting effect of the flow on the pressure gage. When inspector's test connection at top of system is operated, the effect on pressure is also noted here.

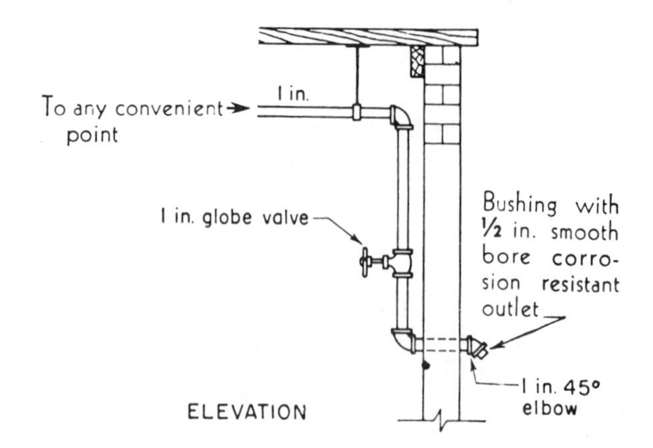

Figure 14-4. Test outlet near top of a sprinkler system to provide a flow approximately that of one sprinkler.

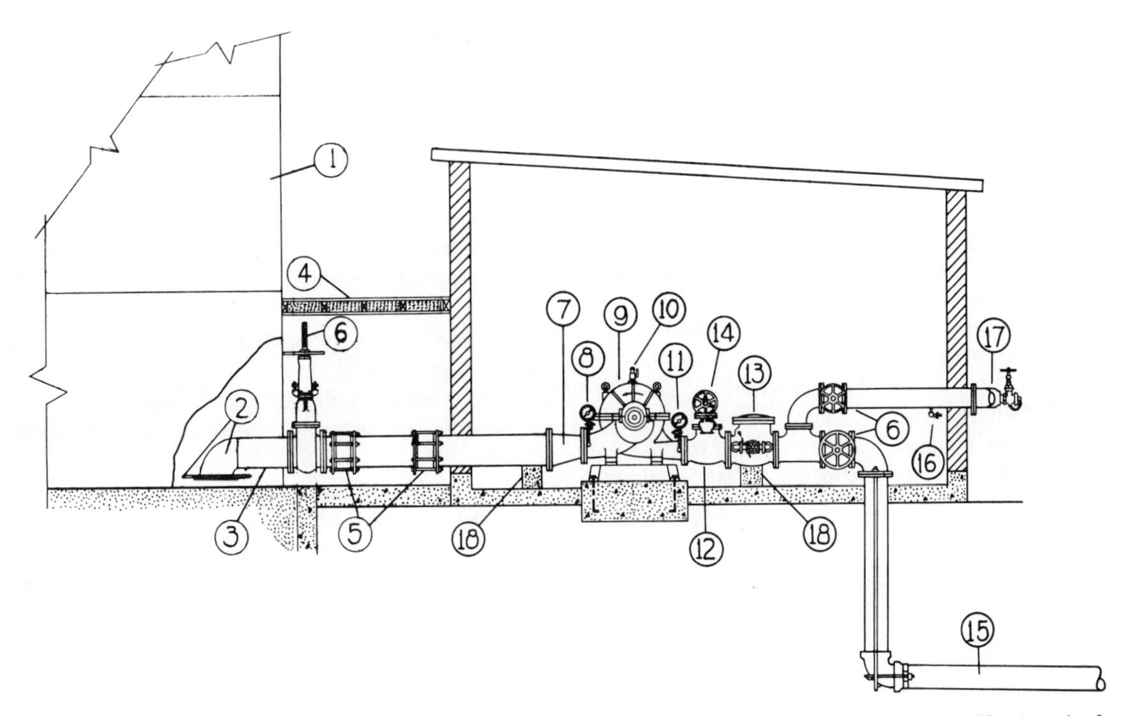

Figure 14-5. Components of a centrifugal fire pump installation where pump always takes water under a head. Numbers in drawing refer to the following: 1. Above ground suction tank. 2. Entrance elbow and vortex plate (vortex plate 4 feet by 4 feet square, 4 inches above bottom of tank). 3. Suction pipe. 4. Frostproof casing. 5. Flexible couplings. 6. Gate valve. 7. Eccentric reducer. 8. Suction gage. 9. Horizontal fire pump. 10. Air release. 11. Discharge gage. 12. Reducing tee. 13. Discharge check valve. 14. Relief valve. 15. Discharge pipe. 16. Drain valve or ball drip. 17. Hose valve manifold with hose valves. 18. Pipe supports.

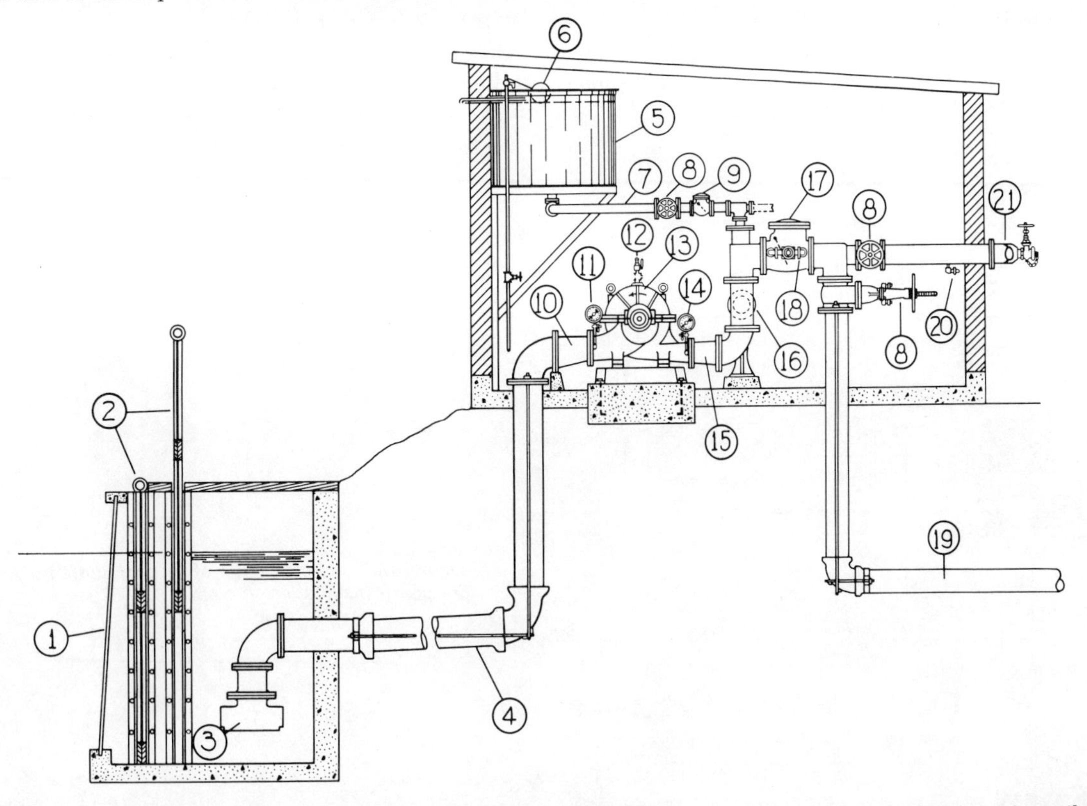

Figure 14-6. Components of centrifugal fire pump installation where pump has suction lift. Numbers in drawing refer to the following: 1. Trash rack, steel bars, ½ inch flat or ¾ inch round, space 2 to 3 inches part. 2. Double screens. 3. Foot valve. 4. Suction pipe from water supply to pump. 5. Priming tank. 6. Automatic float valve. 7. Priming connection. 8. Gate valves. 9. Priming check valve. 10. Eccentric reducer. 11. Suction gage. 12. Umbrella cock. 13. Horizontal fire pump. 14. Discharge gate. 15. Concentric reducer. 16. Relief valve. 17. Discharge check valve. 18. Priming bypass. 19. Discharge pipe. 20. Drain valve or ball drip. 21. Hose valve manifold with hose valves.

(*f*) Where there is central station supervision of alarm valves, in general notify the central station before operating any valve or otherwise disturbing the sprinkler system.

Inspecting Sprinkler Dry-Pipe Valves

Sprinkler systems that must keep the piping filled with air because of danger of freezing have a rather complicated dry-pipe valve, which must be taken care of in accordance with the manufacturer's instructions. The operation and servicing of these valves should be handled only by experienced people. Instruction charts posted near the valves should be consulted. This advice applies particularly to thermostatically controlled dry-pipe and deluge systems. Some of the more important points to be covered in an inspection follow:

(*a*) See that the air pressure on the system is at the required level at least once a week, and pump up the system when necessary. It is necessary to watch the maintenance of air pressure on dry-pipe systems carefully.

(*b*) See that priming water is maintained at the proper level above the dry-pipe valve.

(*c*) In cold weather, see that the valve closet is properly heated and that the heating equipment is in order.

(*d*) Before and during freezing weather, see that all low point drains of the dry-pipe system are free of water.

(*e*) Once a year during the warm weather, it is customary to thoroughly clean and reset each dry-pipe valve. See that this has been done within a year.

Inspecting Fire Pumps

Fire pumps are generally designed for operation only in connection with fire protection service. There are other pumps provided for water needed for other uses in the plant. Some of the points to be covered in an inspection of a fire pump follow:

(*a*) Observe the pump running at its rated speed with water discharging. Check the supply of lubricating oil, the operating condition of the relief valve, and the level of water in any priming tank.

(*b*) In the case of centrifugal pumps, note the condition and reliability of the electric power supply. If the pump is gasoline or diesel engine-driven, examine the storage batteries, lubrication system, oil and fuel supplies.

(*c*) See that the pump room is clean, accessible at all times, and kept at a proper temperature to prevent freezing.

(*d*) See that the suction pipes, intakes, foot valves, and screens of the fire pumps are free from ice or other obstructions.

Once a year, during warm weather, each fire pump should be tested at capacity with hose streams. This provides an excellent exercise for the members of the fire brigade.

Inspecting Water Tanks

A tank is often provided as a primary source of water for fire protection, and it is, therefore, of great importance that it be maintained properly. The in-

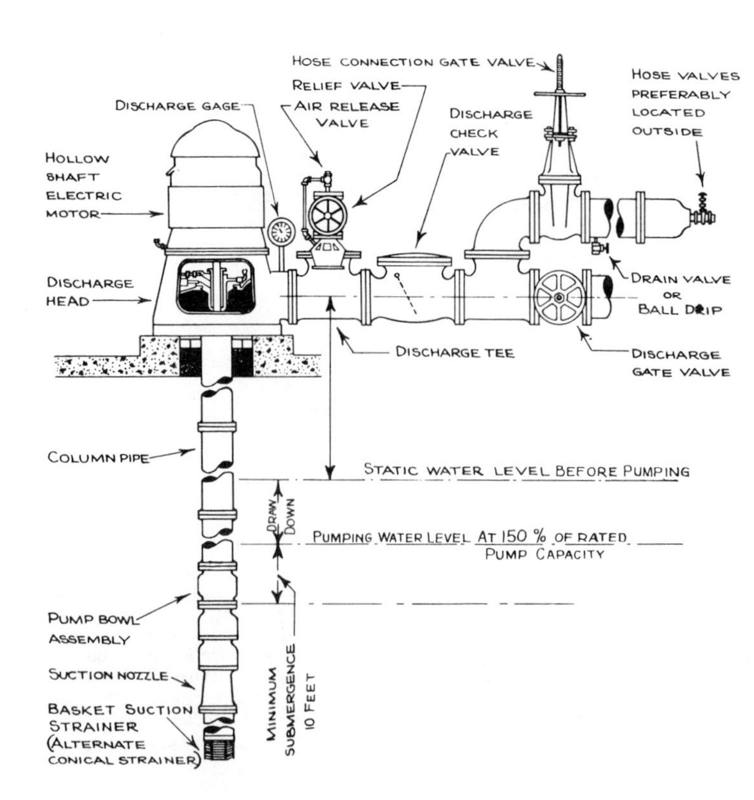

Figure 14-7. Components of vertical shaft turbine-type fire pump installation in a well.

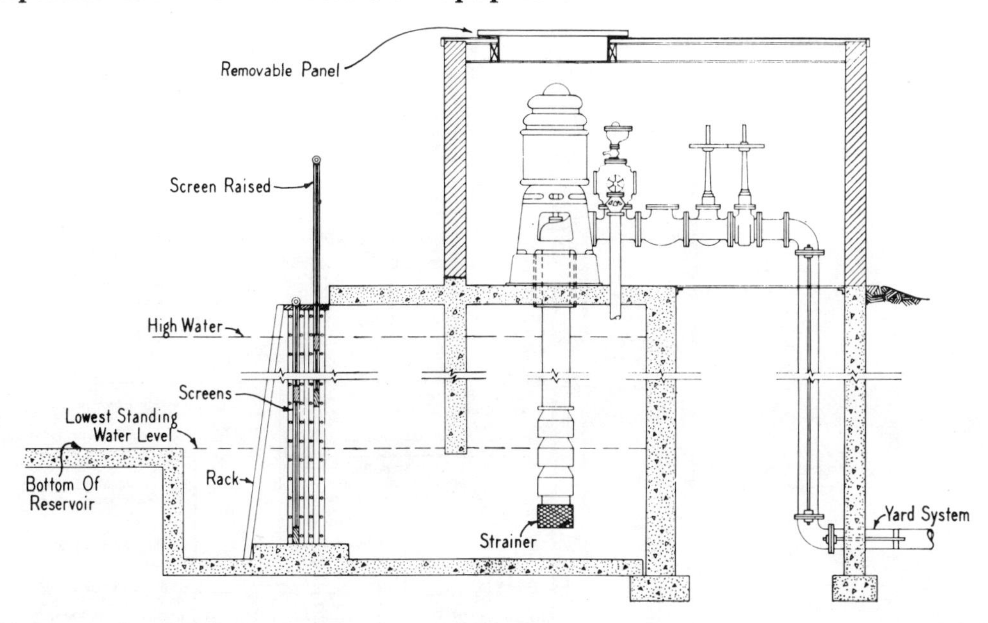

Figure 14-8. Components of vertical shaft turbine-type fire pump installation in a wet pit. When water supply level is below pump level, this arrangement is preferable to the installation shown in Figure 14-6.

spection routine should cover the following points:

(*a*) See that water is maintained at the proper level in the tank. Most tanks have a mercury gage, which can easily be read.

(*b*) See that heating devices are in order and read the thermometer, which is usually provided in the drop pipe of the tank, daily during freezing weather.

(*c*) Observe that the tank roof is kept tight and in good repair, with hatches closed and fastened, and that any frostproof casing of the tank riser is in good repair and makes a tight joint with the bottom of the tank. Freezing in the riser may cut off the supply of water, while the formation of a layer of ice may prevent flow from the tank. The formation of heavy icicles through leaking of the tank is dangerous.

(*d*) See that the space at the top of the tank, the valve pit at the bottom of the tank riser, and the entire area about the bases of the columns of the tank are kept free of dirt, rubbish and waste material.

(*e*) See that the steel work of steel tanks, and hoops of wooden tanks, and the structures of supporting towers are kept properly painted. Examination for this point should be made carefully at least once a year.

(*f*) In the case of pressure tanks, check the water level, air pressure, and, during freezing weather, the heating of the tank enclosure.

Inspecting Fire Doors and Shutters

Fire doors and shutters are provided to retard or prevent the spread of fire from one building to another. The inspection of fire doors and other cut-offs should cover the following items:

(*a*) Note if there are any openings in fire walls not properly protected by fire doors or shutters.

(*b*) See that obstructions, such as stock or machinery, have not been installed so that they prevent the proper manual or automatic closing of the doors.

(*c*) See that fire doors and shutters and their hardware, including fusible links, are in good condition. Paint on fusible links should be scraped off.

(*d*) Test the action of automatic sliding and counterbalanced doors by raising the counterweight by hand.

(*e*) Check the guides on rolling steel doors. These guides sometimes become so badly damaged by trucks that the door cannot drop the entire distance. Operate the doors occasionally by disconnecting the fusible links.

CARE OF FIRE BRIGADE EQUIPMENT

Members of the fire brigade at most plants are specifically charged with the care of equipment the brigade itself will use. This includes hose, ladders, salvage covers, and other tools and equipment. In large plants with automobile fire apparatus, it may sometimes include care of a pumper or other trucks and upkeep of a fire station.

Keeping fire equipment stations and fire fighting apparatus in good condition is principally a matter of having a satisfactory routine. Some things that will help are the following:

(*a*) A daily routine for cleaning the station.

(*b*) A company record book and a check list of items to be reported on daily, weekly, or monthly.

(*c*) Checking fire fighting vehicles and pumps as to gasoline, oil, tires, battery, radiator, condition of hose and tools after each run.

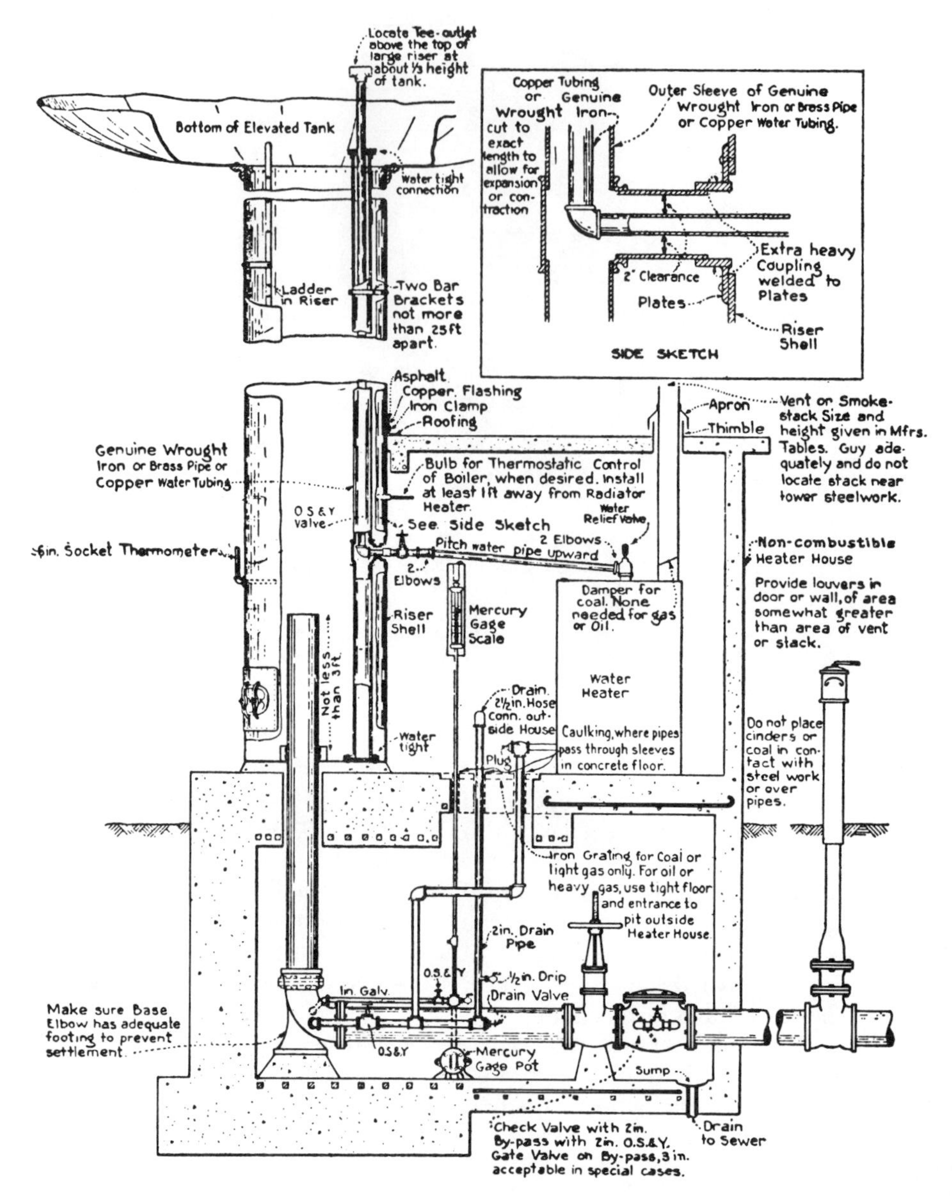

Figure 14-9. Components associated with a gravity tank are illustrated. These will differ in various cases depending on the tank equipment provided. This shows a tank with valve pit at base of a large drop pipe and housing for a water heater. Members of the fire brigade should be familiar with the tank, valves, and accessories at the property they serve.

(*d*) Daily inspections by the responsible officer of each fire company, its station, equipment, and apparatus.

Fire Hose

Successful fire fighting depends to a great extent upon adequate fire streams, and adequate fire streams depend upon well-made and properly maintained fire hose. When hose has been neglected for one reason or another and fails during a fire, time is lost in replacing it. This delay can cause extra loss of property and endanger human life. In order to be sure of the ability of fire hose to stand up under the stress of severe fire conditions, it is necessary to do three things:

(*a*) To purchase properly made hose.

(*b*) To protect the hose during use and storage.

(*c*) To give the hose frequent inspection.

Many purchasing agents buy fire hose (including hose with jackets treated to resist mildew) labeled by Underwriters Laboratories Inc., Underwriters' Laboratories of Canada, or approved by Factory Mutual Laboratories. Purchasers who buy labeled fire hose may, for example, obtain detailed reports of tests made on fire hose by Underwriters Laboratories Inc. Fire hose made, sold, and used without the inspection and label service of the Underwriters Laboratories Inc. or the approval of Factory Mutual Laboratories may be made under the same standard and, when properly tested, prove of equal quality. However, very few

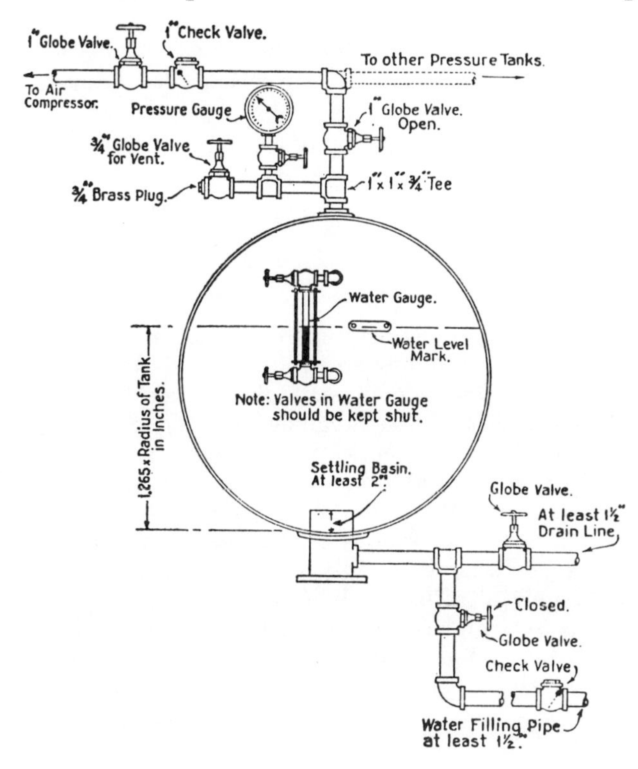

Figure 14-10. Pressure tank diagram showing air piping (at top) and water piping (at bottom).

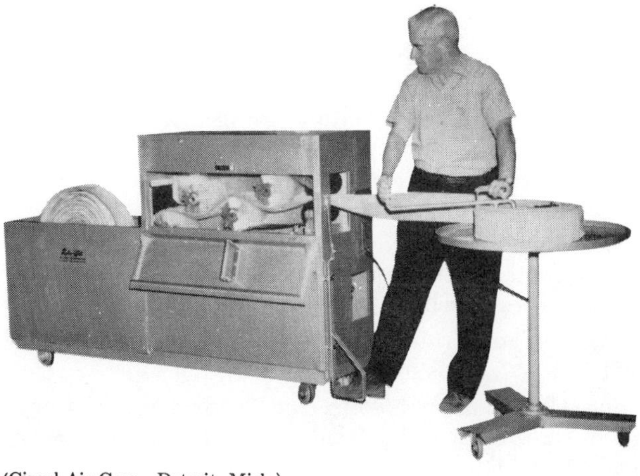

(Circul-Air Corp., Detroit, Mich.)

Figure 14-11. Machine for washing and scrubbing hose.

purchasers are in a position to judge correctly that an order of fire hose comes up to the standard, or are able to conduct adequate tests or make reliable determinations, after delivery, of the quality and workmanship of hose.

Washing and Drying Hose

After hose with woven jackets has been in use at fires, all dirt should be thoroughly brushed off. If the dirt cannot be removed by brushing, the hose should be washed with plain water, scrubbed, and rinsed. It should then be hung up or placed on a rack to dry. The drying should be carefully done. Hose should not be dried in the sun on paved roadways or sidewalks. Electrically controlled hose drying cabinets are efficient and economical.

Hose drying towers are more costly but have been provided at many fire stations. A hose drying rack may be employed where space is available and a drying cabinet is not provided.

Rubber covered hose should be wiped clean and then rewound on a reel or replaced in a hose basket.

Preventing Mechanical Injury to Fire Hose

Common injuries hose receives in use are:

(*a*) Worn places and rips.

(*b*) Abrasions.

(*c*) Crushed couplings.

(*d*) Cracked inner lining.

To prevent this damage:

(*a*) Avoid laying hose over rough, sharp corners.

(*b*) Avoid shutting off the nozzle too abruptly, causing water hammer.

(*c*) Avoid dragging couplings on concrete or hard surfaces. Swivels can be dented by rough usage, and threads are easily damaged. Dragging injuries are less likely when the couplings are connected, because the threaded end is protected by the swivel.

(*d*) Do not permit vehicles to run over fire hose or couplings. Provide warning lights, and use hose bridges in traffic lanes.

(*e*) Change position of bends in hose when reloading.

(Circul-Air Corp., Detroit, Mich.)

Figure 14-12. Cabinet for drying hose.

Preventing Damage to Hose from Heat

Excessive heat resulting from contact with fire will char and weaken woven fabric and dry the rubber lining, making the hose unsuitable for use. The same condition may occur in inner linings when hose is hung in a drying tower for too long a period in high temperatures or when kept for long periods in a heated room. To prevent this damage:

(*a*) Protect hose from excessive heat, firebrands, or fire wherever possible.

(*b*) Use moderate temperatures for drying. A current of warm air is much better than hot air.

(*c*) Do not allow hose to continue to hang in a drying tower after it is dry. Take it down. Roll it up and place it on edge on a storage rack.

(*d*) Keep the woven jackets of fire hose dry.

(*e*) Take hose from hose houses each month and run water through it. Do the same with hose on fire trucks not used within a month. The inside rubber lining benefits by having water run through it periodically.

Preventing Mildew and Mold Damage to Hose

Mildew and mold can develop on the woven jackets of fire hose where moisture is allowed to remain on the jackets for a prolonged period. This leads to decay of the woven fibers.

The handling of hose periodically provides opportunity for inspection, cleaning, wetting the lining, and drying the hose. This should be done even where hose has a synthetic jacket not subject to mildew. Hose being cleaned and dried should be replaced in hose houses and fire trucks with fresh hose.

Keeping Hose Away from Chemicals and Oils

Many liquids and gases contain chemical ingredients that are injurious to hose. Gasoline, if allowed to come in contact with hose, will penetrate the woven jacket and will have a solvent action on the rubber lining. Acid used in soda-acid extinguishers and in storage batteries will also injure hose on contact.

Following are some recommended good practices:

(*a*) Wash the hose with soap when it has been exposed to oil.

(*b*) Guard against spilling of gasoline on hose.

(*c*) Thoroughly scrub and brush all traces of acid contacts with a baking soda solution. The soda neutralizes the acid.

Unlined Linen Fire Hose

Unlined hose is distinguished from rubber-lined hose by the fact that it has no rubber lining. It is used mostly for installation at standpipes inside buildings and similar places where it can be kept dry. Such hose is generally put into service in buildings without subjecting it to any water since, once used and wet, the performance of unlined linen hose is somewhat doubtful, even though much care is used in drying it out. It tends to deteriorate rapidly if not thoroughly dried immediately after use or if installed where it will be exposed to dampness or the weather. It is not built to withstand frequent service.

Linen hose must not be permitted to come in contact with oil or grease or with any corrosive chemicals, such as battery acid. It should be stored in places where rats or other rodents cannot cause injury. Spare hose should be stored in dry rooms in such a way that air will circulate around it. It should never be stored in contact with damp floors or walls.

As a general rule, unlined hose should never be wet except in the event of fire. As it deteriorates rapidly when slightly moist or damp, this prevents hose from being tested hydrostatically (under water pressure). It should, however, be periodically examined to determine if moisture has found its way into the hose or if it is in a satisfactory condition.

Linen hose that has been wet can be dried satisfactorily and replaced in service if careful precautions are taken to assure complete drying. This is so difficult to

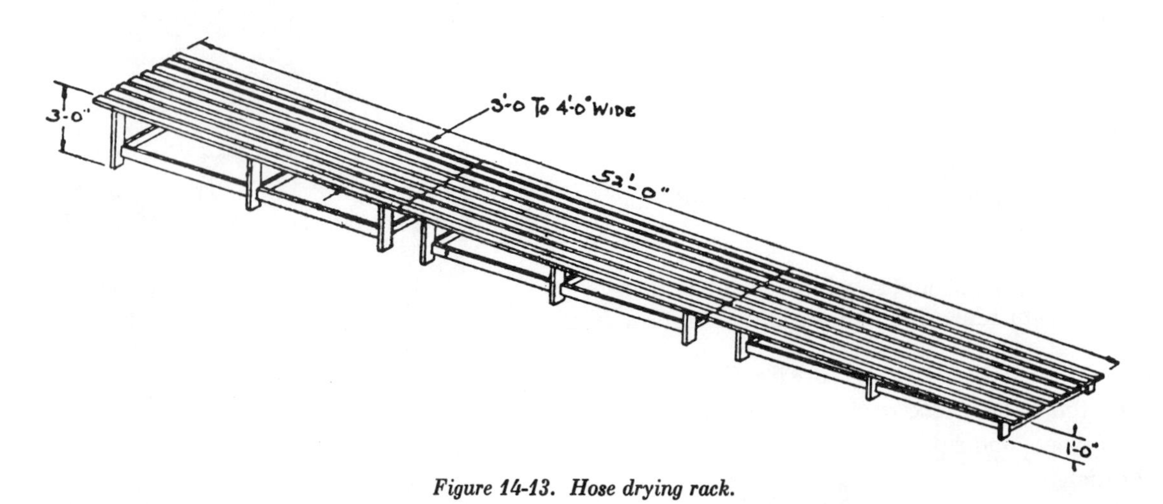

Figure 14-13. Hose drying rack.

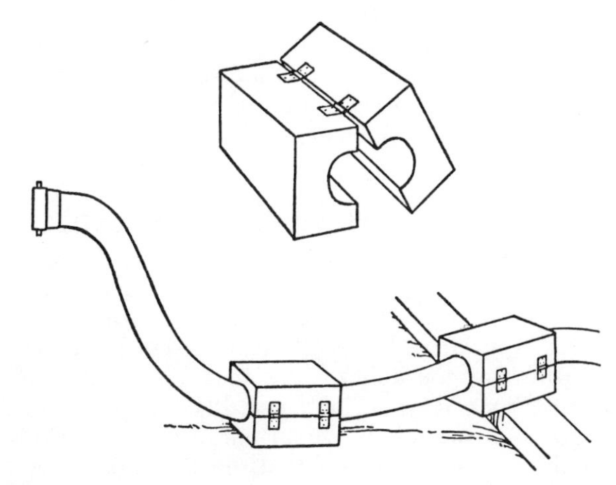

Figure 14-14. Chafing blocks for the protection of hose, particularly pumper suction hose.

do properly that, in general, unlined linen hose should be simply replaced with new hose when used.

Ladders

Fire brigade ladders may receive a great deal of hard usage. Since a ladder that fails in service may mean death or severe injury, it is important that all ladders be subjected to frequent inspection. The inspection could be made conveniently when the ladders are used for weekly drilling.

Inspection should cover all parts of the ladder. It is important that ladders be kept clean to avoid the abrasive action of dirt and grit. Those parts that receive the greatest wear should be given particular attention. Trusses and beams should be checked to determine whether they are cracked or splintered. Rungs are likely to be bruised at the point where they come in contact with the pawls of the fly sections. If there is the slightest doubt as to the safe condition of the ladder, it should be placed out of service until an appropriate load test can be made.

When rungs or beams of wooden ladders are found to be splintered or damaged, the injured part, if it does not require replacement, should be smoothed and sanded immediately, and one coat of filler and two coats of varnish should be applied for protection. It is important that injuries of wooden parts be corrected before the ladder is used again. If even a small amount of moisture penetrates into the wood the whole structure of the ladder will be weakened.

The metal parts of both wooden and metal ladders should also be examined. Aluminum ladders that have been subjected to flame may have been weakened. If the metal has been discolored, the ladder should be tested with weights suitable for the size ladder in question. Fly ladder stops sometimes are worn sufficiently to permit the fly to pass beyond the stop. Tie rods of wooden ladders may become loose through constant

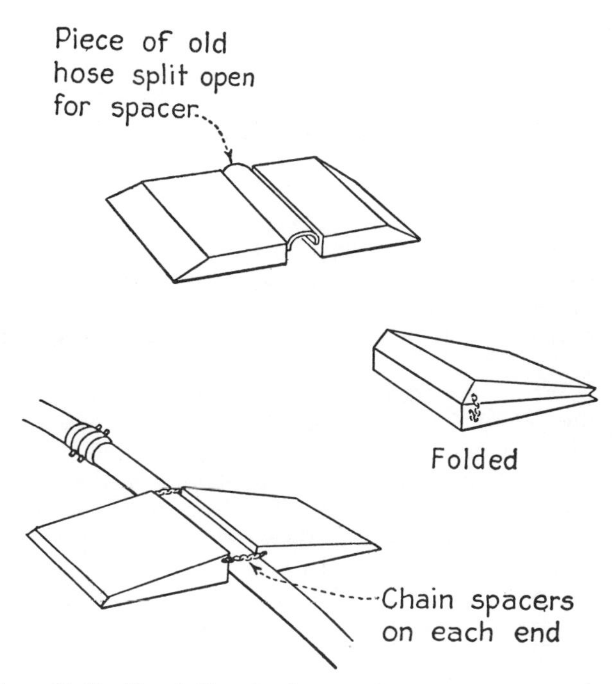

Figure 14-15. Hose bridges for the protection of hose that must be laid in streets.

stress and vibration. Iron or steel parts should be kept painted to prevent rusting. Pulleys and dogs should be oiled during the weekly inspection.

The halyard rope or cables should be examined for cut, worn, or frayed places. Sometimes the end of the halyard that carries the weight of the fly ladder is fastened to a harness snap, which, in turn, is fastened to the fly. This arrangement may not be as safe as an eye spliced directly to the metal support on the fly ladder. Ropes should be replaced when dry or brittle.

Salvage Covers

All salvage equipment needs care and upkeep. However, the salvage equipment most susceptible to injury is the salvage cover. Every fire brigade should have procedures governing the care of covers. Some practical suggestions follow:

(*a*) Covers removed from buildings and places where they have been used should be folded before being conveyed to quarters.

(*b*) The throwing of covers from roofs, windows, or other above-grade locations should be strictly prohibited.

(*c*) When removing covers from stock or machinery, take care to avoid pulling covers over sharp projections to prevent tearing.

(*d*) Covers should be washed and hung on drying racks as soon as possible after being returned from fires.

(*e*) After covers are dry, they must be examined for holes and patched before they are refolded.

(*f*) If not used within a period of one month, covers should be washed if necessary and refolded.

The best method to use in repairing tears and holes in canvas covers is to sew them by hand with a very close "herringbone" stitch, covering the complete job with beeswax and pounding lightly with a wooden mallet. The beeswax should also be used on the thread. Patches should be put on with a double seam by hand or machine. A salvage cover repair kit should be kept on hand at all times. This can be made from any small box and be equipped with the necessary tools, such as thread from the local shoe shop, a chunk of beeswax, an assortment of medium-sized needles, and scissors.

After covers have been used at fires, whether or not they seem dirty, they should be washed and dried. Clear water and a broom usually are all that is necessary, but washing powder is sometimes used when excess dirt is present. Where "wet" or treated water has been used in fire fighting, the wetting agent or penetrant may affect the ability of the cover to shed water, and thorough washing should be employed to remove all traces of the chemical.

Small Tools

"A place for everything, and everything in its place," is the first thing that needs to be said on this subject. Tools should be easy to get when needed.

Axe handles should be kept smooth and oiled with hot linseed oil at least once a year to keep them in good shape and appearance. Cutting tools should be kept serviceably sharp.

Pike poles should be cared for in the same manner as ladders.

Iron tools should be wiped dry to prevent rusting. Adjustable tools need to be kept lubricated to make them work easily.

Automotive Fire Apparatus

If the brigade has automobile fire trucks, they should be refueled, cleaned, and inspected after each run and cleaned and inspected daily between runs.

Tires

Tires should be kept free from mud, grease, and oil. Oil especially is injurious to rubber. If oil or grease gets on the tires, wipe them clean with neutral soap and water. Tires should be kept inflated to the pressure recommended by the manufacturer for the axle loading. A record of inflation often serves as evidence in tire failures.

Batteries

Batteries on fire apparatus should be kept on charge in the station and daily readings of charge should be taken to maintain charge. Care of the battery includes keeping it clean and filled with distilled water. The acid never leaves the battery unless it is overturned or the case is cracked. The water evaporates and must be replaced at intervals. Battery terminals should be checked to see that they are tight, and should be greased to prevent corrosion.

SUPERVISING REPAIRS, CHANGES AND EXTENSIONS

The value of a fire loss prevention organization in a property is nowhere better illustrated than in connection with its responsibility to see that repairs, changes, and extensions do not intefere with fire protection. Changes from summer to winter, for example, institute new daily procedures to see that fire protection is not impaired by freezing.

In many instances where a yard system or a portion of a yard system is shut off, it is possible to temporarily feed the sprinkler systems through hose lines run from city hydrants or from yard hydrants still under pressure to the sprinkler risers, making the connection at the 2-inch (50-millimeter) draw-off valve on the riser.

If any work on inside water piping shows evidence that foreign material has gotten into the pipes, this situation should be investigated and the piping flushed.

Close supervision is necessary, especially when underground mains are being laid or repaired. Following are some of the points to be observed:

(*a*) Before water is shut off for repairs, the work should be laid out and all material made ready.

(*b*) As small an area as possible should be shut off, and someone should be stationed at the closed valves ready to open them if fire breaks out.

(*c*) Extra guards should be provided to patrol rooms where water is shut off.

(*d*) Portable fire extinguishers should be kept ready for instant use.

(*e*) Hose lines should be laid in place for immediate action.

(*f*) Plugs and caps should be provided, so that open pipes can be plugged or capped quickly if the water system must be turned on to deal with a fire.

(*g*) Care should be taken to see that foreign material is kept out of the piping and that joints are properly made.

(*h*) When work is completed, a check should be made to see that all valves are open wide. In addition, drain tests should be made to prove that water pressure has been restored to affected areas.

SERVING AS GUARD

From time to time, members of the fire brigade may be called upon to serve as guards or patrolmen. The patrolman has a job of great responsibility as he is often in sole charge of a plant for more than half the time. Therefore if a person is selected to be a guard, it should be taken as a high compliment to him as one who can be trusted with responsibility.

When serving as a patrolman, the brigade member will probably be given a definite route to cover. This should:

(*a*) Cover the entire area.

(*b*) Preferably not force the guard to retrace steps.

(*c*) Not permit shortcuts by stairways, elevators or bridges.

(*d*) Not require more than 40 minutes.

The patrolman usually is required to report at a stated time shortly before the departure of those whose responsibility he assumes. He begins his first round as soon as the activities of the day or shift stop. He generally carries a reliable electric lantern or flashlight.

The first round is a real inspection of the premises. During this, he should look for conditions that might cause fire and correct them. When he goes on at the close of the day's work, he should make sure that the property is properly closed for the night. Any serious condition, such as a closed sprinkler valve or other impairment to fire protection, should be reported at once to the superintendent or other responsible official.

The patrolman, therefore, has certain definite instructions. These usually cover the following points:

(*a*) The location and use of shut-off valves, drains and alarm devices.

(*b*) To give a fire alarm at once and to summon aid if fire is discovered or if the sprinkler alarms operate. Knowing exactly how to give the alarm is a patrolman's first duty. His instructions should be such that giving the alarm provides responding fire fighters, public or private, with specific information as to the location of the fire. Second, he must clearly understand that he must not attempt to extinguish incipient fires without first giving an alarm to the private fire brigade or to the public fire department or both. A chief cause of delayed alarms is the practice of notifying management personnel before notifying the fire fighting force.

Guards can either make or break a plant by the care with which they do their work. Fires have often been due to smoking by guards. There have been many fire losses where a guard failed in one or more of the following particulars, so it is extremely important not to overlook any of these items:

(*a*) Failure to discover important fires promptly.

(*b*) Trying to fight a fire before turning in an alarm.

(*c*) Shutting off sprinklers without finding out whether the fires have been extinguished.

(*d*) Ignorance of proper sprinkler valves to close after the fire is extinguished.

(*e*) Lack of knowledge as to whom to call to get salvage work started.

Suggested Reading

Fire Protection Handbook, 14th Edition, 1976, NFPA, Boston.

NFPA Inspection Manual, 4th Edition, 1976, NFPA, Boston.

Ely, Robert, *A Fire Officer's Guide, Fire Apparatus Maintenance*, 2nd Edition, 1975, NFPA, Boston.

NFPA 13A, *Recommended Practice for the Care and Maintenance of Sprinkler Systems*, 1976, NFPA, Boston.

NFPA 14, *Standard for the Installation of Standpipe and Hose Systems*, 1976, NFPA, Boston.

NFPA 20, *Standard for the Installation of Centrifugal Fire Pumps*, 1976, NFPA, Boston.

NFPA 22, *Standard for Water Tanks for Private Fire Protection*, 1976, NFPA, Boston.

NFPA 26, *Recommended Practices for the Supervision of Valves Controlling Water Supplies for Fire Protection*, 1976, NFPA, Boston.

NFPA 27, *Recommendations for Organization, Training and Equipment of Private Fire Brigades*, 1975, NFPA, Boston.

NFPA 198, *Standard for Care, Maintenance and Use of Fire Hose*, 1972, NFPA, Boston.

NFPA 601, *Recommendations for Guard Service in Fire Loss Prevention*, 1975, NFPA, Boston.

NFPA 601A, *Standard for Guard Operations in Fire Loss Prevention*, 1975, NFPA, Boston.

NFPA 1901, *Standard for Automotive Fire Apparatus*, 1975, NFPA, Boston.

NFPA 1931, *Standard on Fire Department Ground Ladders*, 1975, NFPA, Boston.

Chapter 15

PRODUCING EFFECTIVE FIRE STREAMS

This chapter is about the streams of water used in fire fighting. Water is the best extinguishing agent readily available for fire fighting. Industrial fire fighters should improve their ability to choose the right fire stream for the particular fire and be able to produce that stream when it is needed.

Most fires are attacked while only a relatively small amount of combustible material is burning. A fire fighter will use small streams much more often than large ones. When the amount of material burning is large, however, large streams are required. Since there will be few chances to get experience in using large streams at actual fires, time must be devoted to the development of large streams in practice sessions. Only then can fire fighters become proficient in producing such streams when needed and in the various methods by which streams can be delivered on a fire at the point where they are wanted.

FIRE STREAMS

There are four general categories of fire streams:

(*a*) Streams supplied through ¾-inch (19-millimeter) or 1-inch (25-millimeter) hose. These are usually the first-aid or "booster" lines from a water tank on fire apparatus. Those streams represent small flows, seldom as high as 30 gallons (114 liters) per minute.

(*b*) Streams supplied through 1½-inch (37-millimeter) hose. These may be supplied from standpipe outlets in buildings, hydrants, or from automobile fire apparatus. A common arrangement is two "leader lines" of 1½-inch (37-millimeter) hose supplied from a 2½-inch (62-millimeter) hose line from pumper or hydrant. Two fire fighters can usually handle the lines of 1½-inch (37-millimeter) hose. Fire fighters will more often use 1½-inch (37-millimeter) lines than lines of other sizes. These streams may flow up to 125 gallons (473 liters) per minute.

(*c*) Streams supplied through single lines of 2½-inch (62-millimeter) hose. These streams are used where relatively large amounts of water are needed, are heavy, and are not mobile in the sense that the smaller lines are able to be quickly moved from place to place. Two or more fire fighters are usually needed to direct the nozzle. Often such lines are only set up where the nozzle can be mounted or lashed in place and the stream directed on one part of the fire for some minutes at a time. These streams represent flows of the order of 175 to 325 gallons (662 to 1,230 liters) per minute.

(*d*) Large streams, which are directed through large nozzles and which are usually supplied through several lines of 2½-inch (62-millimeter) hose. Commonly, two or more lines are siamesed into a deluge set or portable monitor nozzle assembly. In industrial plants where lumber or other large piles of combustible material are stored in a defined yard storage area, it is common to find that arrangements for producing these large streams are permanently installed. Such arrangements include permanent monitors, sometimes mounted on towers, so that they can sweep the storage area. Also it is common to supply the water for these streams through a system of permanently installed piping, so that it is not necessary to lay hose to put these streams in service. Waterfront areas are sometimes served with monitor nozzles from fireboats. These large streams are referred to as monitor nozzle streams or in some fire departments as master streams.

FIRE STREAM PATTERN

The pattern of a fire stream is determined by the fire conditions with which the fire fighters are trying to deal. Water will produce a cooling effect, but the successful transfer of this cooling effect into fire extinguishment is the objective of producing a fire stream.

Water cools by absorbing heat, and it is one of the best substances for absorbing heat. A British thermal unit, which is one of the measures of heat, by definition is the amount of heat absorbed by one pound of water as the water is raised one degree Fahrenheit.

Since 8.3 pounds (3.8 kilograms) of water is one U.S. gallon, water absorbs 8.3 British thermal units (8.8 kilojoules) for each degree Fahrenheit it is heated.

If water from a hose stream is put on a fire at 60 degrees Fahrenheit (16 degrees Celsius) and it runs off at 80 degrees Fahrenheit (27 degrees Celsius), its cooling effect is 20 times 8.3, equals 166 British thermal units (175 kilojoules) per gallon.

If water from a hose stream is raised all the way from 60 degrees to 212 degrees Fahrenheit (16 to 100 degrees Celsius), just short of the boiling point, its cooling effect per gallon is 1,260 British thermal units (1,329 kilojoules).

But if the water is converted to steam, it has great additional cooling effect: 971 (latent heat of water in British thermal units per pound) times 8.3, equals 8,060 British thermal units (8,503 kilojoules) per gallon.

It can be seen, therefore, that, if much of the water is running off after being heated only 20 degrees Fahrenheit, only 166 British thermal units (175 kilojoules) are being absorbed, but if the water is heated to the boiling point *and* converted to steam, 1,260 plus 8,060 or 9,320 British thermal units (9,833 kilojoules) are absorbed.

This arithmetic explains why it is desirable to convert the cooling water from a hose stream to steam if maximum cooling effect is to be achieved. Water applied in the form of a spray or fog is more readily converted to steam. This is one reason why fog or spray streams more effectively use water than the same amount of water used in a solid stream pattern.

A given amount of water applied has only so much cooling effect. If not enough water is used, the fire cannot be put out. A spray pattern may utilize a greater proportion of the cooling effect available. Problems in getting the right total amount of water for fire streams are discussed a little later in this chapter.

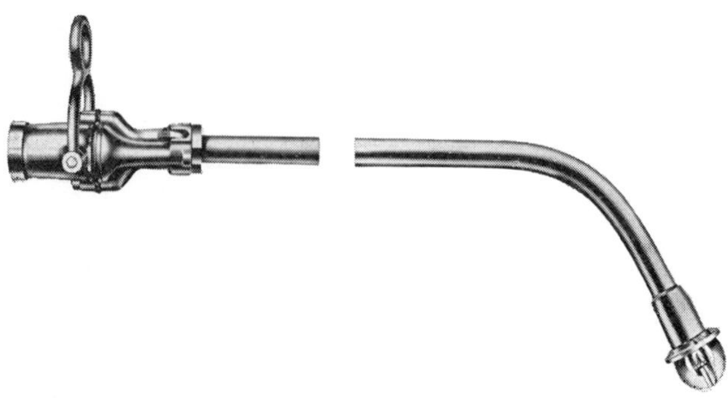

(Grinnell Co., Providence, R.I.)

Figure 15-1. Nonclogging spray sprinkler type spray applicator attached to a combination spray and straight stream nozzle. Applicators are available for a variety of flows and patterns.

Wide-Angle Sprays

Wide-angle sprays are theoretically the most efficient from the point of view of the use of the cooling effects of the amount of water available. They break the water up into relatively small particles. The many small particles present a greater total surface area of water to the combustible material than does a solid stream. This allows a greater proportion of heat to be transferred, with resultant better cooling. Wide-angle sprays are useful for another practical reason. They tend to shield the person using the spray stream. That very advantage produces a disadvantage. A wide-angle spray requires that the nozzle be brought quite close to the burning material. The reach of wide-angle sprays may be increased by the use of applicators, pipes usually 8 to 12 feet (2.5 to 3.6 meters) long, with the nozzle at the end.

Narrow-Angle Sprays

Spray streams may be given a variety of characteristics according to the nozzle design. A flat or fan-shaped stream can be produced by the use of an orifice or nozzle that is essentially a slit. A cylindrical orifice (a straight stream nozzle) with a simple deflector in the stream a short distance from the orifice produces a spray in the shape of a hollow cone (the familiar garden hose nozzle spray). A more complicated form of deflector will break the water up in such a way that it flows in essentially a solid cone of spray. By the shape of the nozzle, by deflectors, by impinging streams, or by combinations of these methods, a cone can be produced with water particles quite uniformly distributed throughout the cone. Most modern fog nozzles produce this cone-shaped stream.

The wide-angle streams usually have the greatest turbulence in the stream and the smallest water particles at a given discharge. Small particles can, of course, be produced by details of nozzle design with narrow discharge cones. In general, the narrower cones have water particles or droplets of larger individual size which carry through the air better than small ones. Smaller droplets are apt to be carried away by draft and wind currents.

Winds and drafts seriously affect the reach of all fire streams. This will be discussed further as straight streams are considered.

Spray streams have very limited reach against the wind or draft as compared to the reach of solid streams, but they are used effectively in special situations. The following are some examples:

(*a*) A wide-angle cone of fine spray may be applied above the surface of burning liquid in an open tank, using an applicator pipe if needed.

(*b*) In various locations, the fire may be vented at one point and a fog stream introduced at another. The fire draft picks up the fog particles and carries them to the fire; there they deliver their cooling effect.

(*c*) A large volume of spray from a large fog nozzle at the end of an aerial ladder may be allowed to drift downward over a fire.

(*d*) A room or building is briskly afire inside, and a fog nozzle, delivering water at such a rate that it is promptly converted to steam, is brought into play. The conversion of water to steam results in the water's volume increasing approximately 1,650 times as it passes from the liquid state to the vapor state. This not only produces a greater cooling effect, but also tends to smother the burning material by displacing air and combustible gases.

Advantage of such effects as these can be taken in the use of fog streams to make up for limited reach.

As the cone of the fog nozzle is made progressively narrower, eventually the point is reached where it is essentially a straight stream. Note that what we call a straight stream assumes the shape of a narrow cone of spray at some distance from the nozzle.

Straight Streams

For straight streams, the object of nozzle design is not to break the stream at the nozzle, but to shape the stream so that it will hold together and carry the water for some distance.

When straight streams are used in fire fighting, the extinguishing effect is obviously obtained from the spray at the far end of the stream. In using a straight stream, it is so directed that it will break up into small water particles and be converted to steam. There is no extinguishing effect gained by keeping the stream flowing on a spot where its cooling effect cannot be used. In such a case, the water merely runs off to do unnecessary water damage.

The quality of a straight stream is judged by the distance at which it will keep together as a solid stream. This is very much a matter of observation and judgment. One basis of measurement, which has been generally accepted, is to state the reach of a solid stream at the point farthest from the nozzle at which most of the stream is going through a 15-inch (38-centimeter) circle.

(Elkhart Brass Mfg. Co., Elkhart, Ind.)

Figure 15-2. Spreader tip to give fan shaped discharge.

Figure 15-3. First in a series of three pictures showing cone patterns. Nozzle illustrated is an adjustable type, 1½-inch size with cone set approximately 90 degrees, discharging 72 U.S. gallons per minute at 100 pounds nozzle pressure. Flow is greater at wide settings than with narrower cones in this and numerous other makes of fog nozzle, but in some makes and models, maximum flow may be at other settings.

Figure 15-4. The second of a series of three pictures showing cone patterns. Nozzle illustrated is 1½-inch size with cone set approximately 60 degrees, discharging 67 U.S. gallons per minute at 100 pounds nozzle pressure. Reach of fog increases as cone is narrowed.

Figure 15-5. Third of a series of three pictures showing cone patterns. Nozzle illustrated is 1½-inch size with cone set approximately 30 degrees, discharging 59 U.S. gallons per minute at 100 pounds nozzle pressure. Cone can be further narrowed to substantially a straight stream in most adjustable cone fog nozzles.

Water will flow out of a straight pipe in a fairly good straight stream. By trial and error, it has been found that if a nozzle is slightly tapered, it makes the straight stream have the most favorable shape on leaving the nozzle. Most modern straight-stream nozzles are of the smooth cone shape inside.

Any small obstruction in the nozzle acts as a deflector and tends to break up the stream. For example, the stream obtained from a shut-off nozzle is never quite as good as that which can be obtained from a nozzle without the tiny projections which are necessary for the shut-off valve mechanism. The valve mechanism produces a small, but measurable, obstruction, but it is of no great practical importance. The advantage of control that the shut-off gives far outweighs the slight decrease in performance. A carefully made shut-off nozzle would not be likely to have rough projections in the water way, but a nozzle in poor condition may seriously affect the kind of stream obtained. A battered nozzle tip can cause too much spray at the nozzle. A very common difficulty is from a poorly fitting gasket at the playpipe or nozzle tip connection.

REACH OF STRAIGHT STREAMS

The nozzle velocity in a stream starts it toward the fire. As soon as the water leaves the nozzle, gravity immediately starts to make the water fall. Therefore, the nozzle velocity must be sufficient to carry the water to the fire before the force of gravity has carried the water to the ground.

If these were the only forces at work, it would be sufficient to increase nozzle pressures to give increased nozzle velocities when fire streams must reach farther. However, another force is exerted by the friction of the air at the outside edges of the stream. This causes the outer water droplets to peel off and be blown away. At some distance, they are all blown away and the stream ceases to deliver any water at all.

The circumference of the nozzle orifice is directly proportional to the nozzle diameter. The circumference of a circle is simply its diameter multiplied by 3.14.

These facts have practical application to the business of getting a good fire stream. Their values may be shown by following through the hydraulic calculations for an example or, better, by trying out actual streams at a practice session.

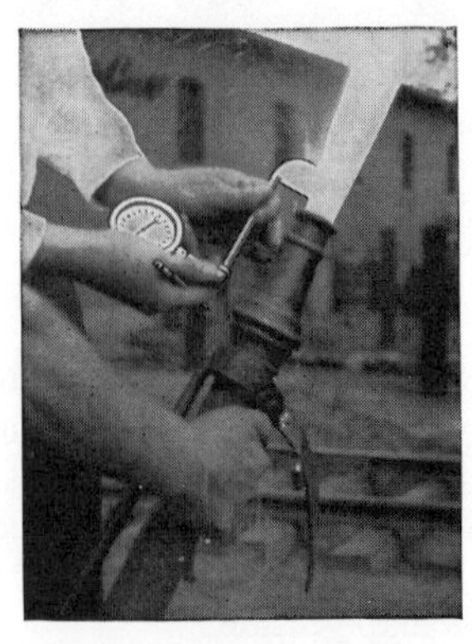

Figure 15-6. Measuring nozzle pressure with a Pitot gage. A Pitot gage assembly is also illustrated in Figure 15-7. Another method is to install a pressure gage in a coupling just back of the nozzle.

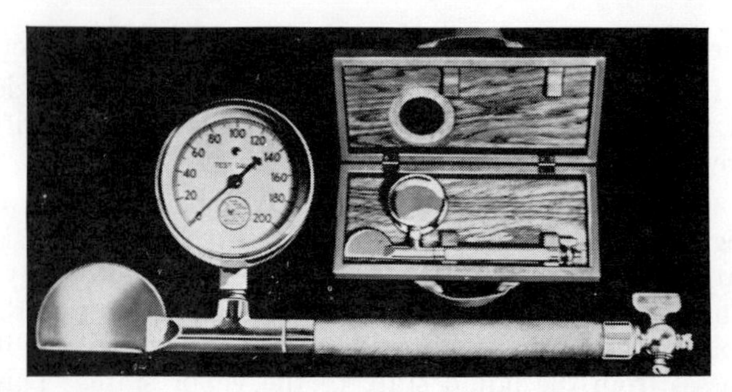

Figure 15-7. A Pitot gage assembly is useful for measuring flows in straight stream nozzles. Velocity of the moving water is picked up at an opening to a small tube at the tip of the knife blade.

Here is a layout useful to illustrate the factors involved.

Lay out a 2½-inch (62-millimeter) hose line from a hydrant. Start with a 1⅛-inch (29-millimeter) tip and 30 pounds (207 kiloPascals) nozzle pressure. The nozzle pressure desired may be obtained by varying the amount of hose used between the hydrant and the nozzle, or by throttling down the supply with a valve at the hydrant connection.

Measure the nozzle pressure with a Pitot gage in the stream itself. This is a convenient and reasonably accurate method of reading nozzle pressure. When the flow is adjusted to 30 pounds (207 kiloPascals) pressure, take a look at the stream.

The amount of water flowing is approximately 200 U.S. gallons (757 liters) of water per minute.

Has the stream a good reach? It will have about 50 feet (15 meters) of effective reach in still air, and it would be classed as no better than a fair stream.

The problem is this. Without changing the flow or pressure at the supply, how can a stream of better reach be obtained?

Close the nozzle shut-off and replace the 1⅛-inch (29-millimeter) tip with a tip of 1-inch (25-millimeter) diameter.

When the water is turned on again the reach of the stream is 10 to 20 feet (3 to 6 meters) better. The amount of water flowing must be the same. A Pitot gage reading at the nozzle will be 50 pounds (345 kiloPascals). However, if the pressure at the 1⅛-inch (29-millimeter) tip is increased to 50 pounds (345 kiloPascals), the stream will reach farther, and the nozzle will discharge more water.

This kind of exercise is particularly useful for building up experience in judging the reach and general performance of fire streams. Detailed figures about flow, nozzle sizes, and pressures cannot be kept in the head. At a fire, there is no time to make detailed hydraulic calculations. Nevertheless, in this example may be learned the relationship for the pressure and reach and and also the flow in two streams. From the figures in these cases a pretty good guess can be made about the reach to be expected of a stream of a different nozzle size and flow.

WATER REQUIREMENTS OF FIRE STREAMS

The student who has followed to this point has learned that a fire stream with good reach can be obtained with a 1-inch (25-millimeter) nozzle at 50 pounds per square inch (345 kiloPascals) nozzle pressure.

One should next learn something about the discharge from other common nozzles. The recommended nozzles in many cases are combination nozzles from which either a straight stream or fog can be discharged. To understand the combination feature, the performance of straight streams and fog streams must be understood.

It is a good idea to learn thoroughly the figures for a few specific straight stream nozzles. There is no advantage in trying the difficult business of learning figures for all the possible variations of flow with different nozzle sizes and pressures.

Pressure and Flow Relationships

¾-inch and 1-inch Lines: The common ¾-inch (19-millimeter) or 1-inch (25-millimeter) hose line produces a stream from a small nozzle and is not for use where large quantities of water are required. The traditional nozzle for such lines was a ¼-inch (6-millimeter) straight stream nozzle. Fog nozzles are now being employed for fire fighting operations. For comparison with the flows of fog nozzles that would be used on ¾-inch (19-millimeter) or 1-inch (25-millimeter) hose, the discharge of a ¼-inch (6-millimeter) nozzle for a range of pressures is tabulated in this chapter.

One-inch (25-millimeter) lines usually consist of 200 feet (61 meters) of cotton hose with a rubber cover and a rubber lining of approximately 1-inch (25-millimeters) inside diameter. Similar hose with a ¾-inch (19-millimeter) inside diameter is also used occasionally. By limiting nozzles on such small lines to less than 20 gallons (76 liters) per minute discharge rates, it is possible to stay comfortably within the capabilities of ¾-inch (19-millimeter) and 1-inch (25-millimeter) hose.

STRAIGHT STREAM NOZZLE

Nozzle Pressures per Square Inch:	50 Pounds	100 Pounds	150 Pounds
Flow in U.S. Gallons per Minute:			
Nozzle ¼-inch diameter	13	18	22
Flow in Imperial Gallons per Minute:			
Nozzle ¼-inch diameter	10	15	18

REACH OF FIRE STREAMS

Straight-Stream Nozzles
(Under ordinary conditions in still air)

	NOZZLE PRESSURE Per Square Inch	REACH
Nozzle 1-inch diameter	30 pounds	50 feet
	45 pounds	60 feet
	60 pounds	70 feet
Nozzle 1⅛-inch diameter	30 pounds	50 feet
	45 pounds	70 feet
	60 pounds	80 feet
Nozzle 1¼-inch diameter	30 pounds	50 feet
	45 pounds	70 feet
	60 pounds	80 feet

FLOW IN FIRE STREAMS

Straight-Stream Nozzles

Nozzle Pressure per Square Inch:	30 Pounds	50 Pounds	60 Pounds
Flow in U.S. Gallons per Minute:			
Nozzle 1-inch diameter	160	200	230
Nozzle 1⅛-inch diameter	200	250	290
Nozzle 1¼-inch diameter	250	330	360
Flow in Imperial Gallons per Minute:			
Nozzle 1-inch diameter	130	170	190
Nozzle 1⅛-inch diameter	170	220	240
Nozzle 1¼-inch diameter	210	270	300

1½-inch Lines: The common straight stream nozzles for 1½-inch (37-millimeter) lines are the ½-inch (13-millimeter) and the ⅝-inch (16-millimeter) nozzles. Fire streams of good reach can be obtained with about 50 pounds (345 kiloPascals) nozzle pressure.

The use of 1½-inch (37-millimeter) lines of hose as "leader" lines is so common that it is well to study the typical leader line situation. In this, 2½-inch (62-millimeter) hose lines may be used to bring water from the hydrant or the pumper to the fire. At the fire,

STRAIGHT STREAM NOZZLES

Nozzle Pressures per Square Inch:	50 Pounds	100 Pounds	150 Pounds
Flow in U.S. Gallons per Minute:			
Nozzle ½-inch diameter	50	73	90
Nozzle ⅝-inch diameter	80	114	140
Flow in Imperial Gallons per Minute:			
Nozzle ½-inch diameter	40	60	73
Nozzle ⅝-inch diameter	66	95	115

one or two lines of 1½-inch (37-millimeter) hose are taken off of the 2½-inch (62-millimeter) feeder hose, and it is the 1½-inch (37-millimeter) lines that are used by the fire fighters. Such lines are more easily moved than larger hose lines, and at ordinary working nozzle pressures, one person can handle the nozzle.

With flows as much as 100 U.S. gallons (379 liters) per minute, there would be 50 to 60 pounds (347 to 414 kiloPascals) pressure required to overcome friction in each 200-foot leader line of woven-jacketed rubber-lined hose. The advantages of good pressures have to be balanced against the disadvantage that the higher the pressure, the more difficult it is for fire fighters to hold or move the lines.

Brigades using fog nozzles should obtain from the manufacturer of each nozzle a statement of the flows to be expected at various pressures. For the types that are adjustable, that is, those which can be used at a wide spray and progressively adjusted to a narrower cone, this is particularly important. Some adjustments involve discharges larger than those of the ½-inch (13-millimeter) straight nozzle, and it is necessary for the operator of the nozzle to be aware of the added demand. The type of nozzle that has a straight tip and a fog tip, which can be used simultaneously, may call for more than double the usual amount of water.

A brigade should run tests of all its fog nozzles so that it knows what flows to expect at various pressures and so that nozzle operators may know how the flows compare with the usual straight stream nozzles used on 1½-inch (37-millimeter) lines.

In all cases where materially less water is used through a fog nozzle than would be through a straight stream nozzle on the same line, the operator must guard against using so little water that the fire extinguishing effect cannot be achieved. A rough rule is that a fog stream can do as much to extinguish fires under usual conditions with half the water required by the comparable straight stream nozzle.

In industrial plants, 1½-inch (37-millimeter) lines will frequently be used from connections on a sprinkler system. In such cases, the pressures available are fixed by the usual pressures in the water supply system. Lines of 1½-inch (37-millimeter) hose usually should have straight stream nozzles no greater than ½ inch (13 millimeters) in diameter, as larger tips require flows and pressures larger than the water system can comfortably supply.

It is best to realize that 1½-inch (37-millimeter) hose lines will usually be employed at nozzle pressures and flows near the lower limits of effective fire streams. A good set of working values to remember are those of a ½-inch (13-millimeter) straight tip from 50 feet (15 meters) of 1½-inch (37-millimeter) woven-jacketed, rubber-lined hose, with 40 pounds (276 kiloPascals) nozzle pressure. The flow will be approximately 46 U.S. gallons (174 liters) per minute, the reach will be 30 feet (9 meters) or more and the pressure at the standpipe or sprinkler riser can be as low as 50 pounds (345 kiloPascals).

The limiting factor of pressure at the standpipe or sprinkler riser applies also when fog nozzles are employed. A good choice would be a nozzle with a discharge at low pressures — 30 to 35 gallons (114 to 132 liters) per minute perhaps.

2½-inch Lines: The common nozzles for single 2½-inch (62-millimeter) hose lines are the 1-inch, 1⅛-inch, and 1¼-inch (25-, 29-, and 31-millimeter). Nozzles

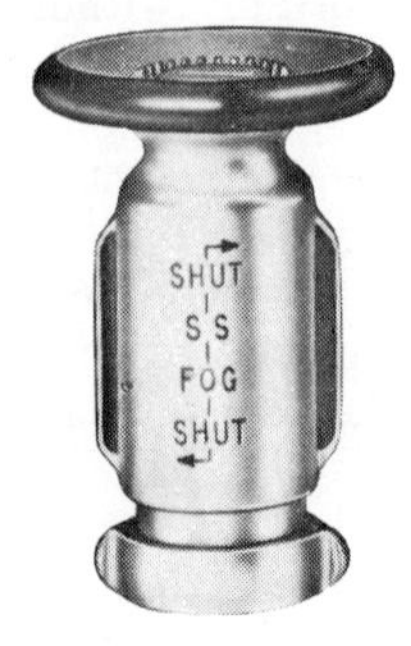

(Akron Brass Co., Wooster, Ohio)

Figure 15-8. A 1½-inch "Akr-O-Jet" adjustable spray nozzle, also available in 1- and 2½-inch sizes.

(American-LaFrance Div. of "Automatic" Sprinkler Corp., of America, Elmira, N.Y.)

Figure 15-9. A 2½-inch "Alcospray" nozzle with adjustable fog cone. This nozzle is also made in 1- and 1½-inch sizes.

¾-inch and ⅞-inch (19- and 22-millimeter) size are sometimes found, but they are skipped here to make the treatment of the subject simpler. When nozzles larger than 1¼-inch (31-millimeter) are employed, it is usually on monitors served by two or more lines of hose. A whole book could be written on hose stream practice, so this chapter cannot be expected to cover it all.

Straight streams from single lines of 2½-inch (62-millimeter) hose are used in the ranges indicated by the two accompanying tables of reach and flow.

For comparison with the flows of fog nozzles, the flows at 50, 100 and 150 pounds (345, 689, and 1,034 kiloPascals) for three straight stream nozzles are tabulated. Pressures of 100 and 150 pounds are too high for hand line use.

There are some spray nozzles that fall in approximately the same range of discharges as each of the three straight nozzles tabulated.

The choice of nozzle size and length of line is made on the basis of fire conditions. The objective is to get enough water on the fire to produce a cooling effect. The amount required is determined by what is burning.

STRAIGHT-STREAM NOZZLES

Nozzle Pressure per Square Inch:	50 Pounds	100 Pounds	150 Pounds
Flow in U.S. Gallons per Minute:			
Nozzle 1-inch diameter	200	300	360
Nozzle 1⅛-inch diameter	250	375	460
Nozzle 1¼-inch diameter	330	460	560
Flow in Imperial Gallons per Minute:			
Nozzle 1-inch diameter	170	250	300
Nozzle 1⅛-inch diameter	220	300	380
Nozzle 1¼-inch diameter	270	380	470

(Akron Brass Co., Wooster, Ohio)

Figure 15-10. "Turbo-jet" 2½-inch fog nozzle adjustable to fog discharge with straight stream and shutoff positions. Also available in 1- and 1½-inch sizes.

The length of line is determined partly by the plant layout and the distance to hydrants.

The amount of water that can be delivered through one line of hose is limited by the amount of pressure

Figure 15-11. First of a series of three pictures showing how fog stream pattern is affected by pressure. Nozzle illustrated is 2½-inch size adjustable cone set approximately 60 degrees, discharging 200 U.S. gallons per minute at 50 pounds nozzle pressure. It shows a good fog even at low pressure.

Figure 15-12. Second of a series of three pictures showing how fog stream pattern is affected by pressure. Nozzle is 2½-inch, 60-degree cone, discharging 300 U.S. gallons per minute at 100 pounds nozzle pressure. Higher pressure not only delivers more water and increases reach of stream, but also greater turbulence produces a better fog.

Figure 15-13. Third of a series of three pictures showing how fog stream pattern is affected by pressure. Nozzle is 2½-inch, 60-degree cone, discharging 360 U.S. gallons per minute at 150 pounds nozzle pressure. More water is being delivered and still greater turbulence produces the smallest fog particles of the three cases illustrated, but air resistance is preventing much additional reach.

HOSE LINES WITH 50 POUNDS HYDRANT PRESSURE

Single lines of 2½-inch cotton, rubber-lined fire hose and pressures in pounds per square inch at hydrant. Nozzle at same elevation as hydrant.

Nozzle Diameter	Length of Line	Nozzle Pressure	Friction Loss
Inches	Feet	Pounds	Pounds
1	300	33	17
1	200	38	12
1	100	43	7
1⅛	200	33	17
1⅛	100	40	10
1⅛	50	44	6
1¼	100	36	14
1¼	50	42	8

HOSE LINES WITH 100 POUNDS PUMPER PRESSURE

Single lines of 2½-inch cotton, rubber-lined fire hose and pressures in pounds per square inch at pumper. Nozzle at same elevation as pumper.

Nozzle Diameter	Length of Line	Nozzle Pressure	Friction Loss
Inches	Feet	Pounds	Pounds
1	600	50	50
1	450	57	43
1	300	67	33
1⅛	450	50	50
1⅛	300	60	40
1⅛	150	75	25
1¼	500	33	67
1¼	250	50	50

used up due to the flow (friction loss). The industrial fire fighter will find it useful to know something about the limits on lengths of individual lines imposed by friction loss.

Where lines are used directly from hydrants, the pressure at the hydrant for a given flow sets a limit on the length of line. Where lines are fed from an automobile pumper, the pressures can be adjusted to the requirements, except that there is a practical top limit to the pressures that can be used. This is usually 175 to 225 pounds per square inch (1,207 to 1,551 kiloPascals). That is the usual pressure range to which woven-jacket, rubber-lined fire hose is subjected in annual tests. Hose in good condition will not burst until subjected to much higher pressures, but the higher the pressures, the greater the risk of failure of hose, couplings, and other equipment being used in the hose line layout. Industrial fire pumps are usually designed to deliver their rated capacity at 100 pounds (689 kiloPascals) and automobile pumpers at 150 pounds (1,034 kiloPascals). Equipment for higher pressures can be provided, of course, but what is said here relates to usual equipment, not that designed for special service.

Industrial fire fighters will find it convenient to know what lengths of hose can be employed with 50 pounds (345 kiloPascals) pressure at the hydrant and at 100 pounds (689 kiloPascals) at the hydrant or pumper. This gives them a knowledge of two points in the pressure range at which they will use hose lines or see them used by a fire department.

Extra High-Pressure Fog Lines

A special type of fog equipment uses pressures higher than those for which ordinary fire equipment is designed. Ordinary equipment is for relatively moderate working pressures with nozzle pressures of the order of 50 pounds (345 kiloPascals) and under 150 pounds (1,034 kiloPascals) in yard piping and hose. Operation at nozzle pressures of 100 pounds (689 kiloPascals) or more with higher pressures in water piping or hose is frequently spoken of as "high-pressure." The special equipment is a piece of motorized fire apparatus with a tank and a pump with rated capacity at 800 pounds (5,516 kiloPascals) and with piping and equipment designed for this exceptionally high pressure. The hose lines are high-pressure hose of ¾-inch (19-millimeter) size.

One manufacturer's model produces a fog stream with a reach of 30 feet (9 meters) and a straight stream of 60-foot (18-meter) reach. Nozzle pressures are 600 pounds per square inch (4,137 kiloPascals) and discharge is 25.5 gallons (96 liters) per minute for the fog stream and 29.5 gallons (112 liters) for the straight stream. On the basis of the amount of water delivered on a fire, note that these "extra high pressure" streams are about three times as large as a stream with a ¼-inch (6-millimeter) nozzle and about half as large as the stream from a ½-inch (13-millimeter) nozzle [as used on 1½-inch (37-millimeter) hose].

A FEW PRACTICAL RULES ON HOSE STREAMS

A few practical rules can be stated and the reasons for each is evident from the foregoing discussions:

(*a*) Use small lines inside buildings because of their greater maneuverability.

(*b*) Use short lines from hydrant supplies on low-pressure yard systems.

(*c*) If water flow is too limited to give a good stream with the nozzle size you have, change to one of smaller size.

(*d*) Make sure the water supply you are using will supply the water for the streams you are trying to use. For example, do not take hose streams from a supply that is already feeding sprinklers unless you are sure the water supply is ample for both.

(*e*) Avoid long lines unless the supply is from pumpers or a high-pressure water system.

(*f*) Have a good idea as to how much water the fire situation requires, and make sure the hose lines and nozzle sizes chosen will deliver it, particularly when using fog or spray nozzles.

(*g*) Use two or more lines in parallel to supply one stream to reduce friction losses when a long line or much water is needed.

Suggested Reading

Fire Protection Handbook, 14th Edition, 1976, NFPA, Boston.

Fire Stream Practices, 5th Edition, 1972, International Fire Service Training Association, Stillwater, Oklahoma.

INDEX

NOTES

NOTES

NOTES

NOTES